RANDOM MATRIX THEORY AND ITS APPLICATIONS

Multivariate Statistics and Wireless Communications

LECTURE NOTES SERIES
Institute for Mathematical Sciences, National University of Singapore

Series Editors: Louis H. Y. Chen and Ser Peow Tan
Institute for Mathematical Sciences
National University of Singapore

ISSN: 1793-0758

Published

Vol. 7 Markov Chain Monte Carlo: Innovations and Applications
edited by W. S. Kendall, F. Liang & J.-S. Wang

Vol. 8 Transition and Turbulence Control
edited by Mohamed Gad-el-Hak & Her Mann Tsai

Vol. 9 Dynamics in Models of Coarsening, Coagulation, Condensation and Quantization
edited by Weizhu Bao & Jian-Guo Liu

Vol. 10 Gabor and Wavelet Frames
edited by Say Song Goh, Amos Ron & Zuowei Shen

Vol. 11 Mathematics and Computation in Imaging Science and Information Processing
edited by Say Song Goh, Amos Ron & Zuowei Shen

Vol. 12 Harmonic Analysis, Group Representations, Automorphic Forms and Invariant Theory — In Honor of Roger E. Howe
edited by Jian-Shu Li, Eng-Chye Tan, Nolan Wallach & Chen-Bo Zhu

Vol. 13 Econometric Forecasting and High-Frequency Data Analysis
edited by Roberto S. Mariano & Yiu-Kuen Tse

Vol. 14 Computational Prospects of Infinity — Part I: Tutorials
edited by Chitat Chong, Qi Feng, Theodore A Slaman, W Hugh Woodin & Yue Yang

Vol. 15 Computational Prospects of Infinity — Part II: Presented Talks
edited by Chitat Chong, Qi Feng, Theodore A Slaman, W Hugh Woodin & Yue Yang

Vol. 16 Mathematical Understanding of Infectious Disease Dynamics
edited by Stefan Ma & Yingcun Xia

Vol. 17 Interface Problems and Methods in Biological and Physical Flows
edited by Boo Cheong Khoo, Zhilin Li & Ping Lin

Vol. 18 Random Matrix Theory and Its Applications
edited by Zhidong Bai, Yang Chen & Ying-Chang Liang

*For the complete list of titles in this series, please go to
http://www.worldscibooks.com/series/LNIMSNUS

Lecture Notes Series, Institute for Mathematical Sciences,
National University of Singapore

RANDOM MATRIX THEORY AND ITS APPLICATIONS

Multivariate Statistics and Wireless Communications

Editors

Zhidong Bai
National University of Singapore, Singapore
and
Northeast Normal University, P. R. China

Yang Chen
Imperial College London, UK

Ying-Chang Liang
Institute for Infocomm Research, Singapore

NEW JERSEY · LONDON · SINGAPORE · BEIJING · SHANGHAI · HONG KONG · TAIPEI · CHENNAI

Published by

World Scientific Publishing Co. Pte. Ltd.
5 Toh Tuck Link, Singapore 596224
USA office: 27 Warren Street, Suite 401-402, Hackensack, NJ 07601
UK office: 57 Shelton Street, Covent Garden, London WC2H 9HE

British Library Cataloguing-in-Publication Data
A catalogue record for this book is available from the British Library.

Lecture Notes Series, Institute for Mathematical Sciences, National University of Singapore
Vol. 18
RANDOM MATRIX THEORY AND ITS APPLICATIONS
Multivariate Statistics and Wireless Communications

ISBN-13 978-981-4273-11-4
ISBN-10 981-4273-11-2

Printed in Singapore.

CONTENTS

FOREWORD

The Institute for Mathematical Sciences at the National University of Singapore was established on 1 July 2000. Its mission is to foster mathematical research, both fundamental and multidisciplinary, particularly research that links mathematics to other disciplines, to nurture the growth of mathematical expertise among research scientists, to train talent for research in the mathematical sciences, and to serve as a platform for research interaction between the scientific community in Singapore and the wider international community.

The Institute organizes thematic programs which last from one month to six months. The theme or themes of a program will generally be of a multidisciplinary nature, chosen from areas at the forefront of current research in the mathematical sciences and their applications.

Generally, for each program there will be tutorial lectures followed by workshops at research level. Notes on these lectures are usually made available to the participants for their immediate benefit during the program. The main objective of the Institute's Lecture Notes Series is to bring these lectures to a wider audience. Occasionally, the Series may also include the proceedings of workshops and expository lectures organized by the Institute.

The World Scientific Publishing Company has kindly agreed to publish the Lecture Notes Series. This Volume, "Random Matrix Theory and Its Applications: Multivariate Statistics and Wireless Communications", is the eighteenth of this Series. We hope that through the regular publication of these lecture notes the Institute will achieve, in part, its objective of promoting research in the mathematical sciences and their applications.

February 2009

Louis H. Y. Chen
Ser Peow Tan
Series Editors

PREFACE

Random matrices first appeared in multivariate statistics with the work of Wishart, Hsu and others in the 1930s and enjoyed tremendous impetus in the 1950s and 1960s due to the important contribution of Dyson, Gaudin, Mehta and Wigner.

The 1990s and beyond saw a resurgent random matrix theory because of the rapid development in low-dimensional string theory.

The next high-water mark involves the discovery of probability laws of the extreme eigenvalues of certain families of large random matrices made by Tracy and Widom. These turned out to be particular solutions of Painlevé equations building on the work of Jimbo, Miwa, Mori, Sato, Mehta, Korepin, Its and others.

The current volume in the IMS series resulting from a workshop held at the Institute for Mathematical Science of the National University of Singapore in 2006 has five extensive lectures on various aspect of random matrix theory and its applications to statistics and wireless communications.

Chapter 1 by Jack Silverstein studies the eigenvalue, in particular, the eigenvalue density of a general class of random matrices — only mild conditions were imposed on the entries — using the Stieltjes transform. This is followed by Chapter 2 of Peter Forrester which deals with those class random matrices where there is an explicit joint probability density of the eigenvalues and the "symmetry" parameter β which describe the logarithmic repulsion between the eigenvalues takes on general values. Chapter 3 by Zhidong Bai is a survey of the future in statistics taking into account of the impact modern high speed computing facilities and storage space. In the next two chapters, one finds applications of random matrix theory to wireless communications typified in the multi-input multi-output situation commonly found, for example, in mobile phones. Chapter 4 by Antonia Tulino uses the Shannon transform — intimately related to the Stieltjes transform discussed in Chapter 1 — to compute quantities of interest in wireless communication. In the last chapter, Ralf Muller made use of the Replica Methods developed by Edwards and Anderson in their investigation of spin-glasses to tackle multiuser problems in wireless communications.

February 2009

Zhidong Bai
National University of Singapore, Singapore
& Northeast Normal University, P. R. China

Yang Chen
Imperial College London, UK

Ying-Chang Liang
Institute of Infocomm Research, Singapore
Editors

THE STIELTJES TRANSFORM AND ITS ROLE IN EIGENVALUE BEHAVIOR OF LARGE DIMENSIONAL RANDOM MATRICES

Jack W. Silverstein

Department of Mathematics
North Carolina State University
Box 8205, Raleigh, North Carolina 27695-8205, USA
E-mail: jack@math.ncsu.edu

These lectures introduce the concept of the Stieltjes transform of a measure, an analytic function which uniquely chacterizes the measure, and its importance to spectral behavior of random matrices.

1. Introduction

Let $\mathcal{M}(\mathbb{R})$ denote the collection of all subprobability distribution functions on $\mathbb{R}$. We say for $\{F_n\} \subset \mathcal{M}(\mathbb{R})$, F_n converges vaguely to $F \in \mathcal{M}(\mathbb{R})$ (written $F_n \xrightarrow{v} F$) if for all $[a,b]$, a,b continuity points of F, $\lim_{n\to\infty} F_n\{[a,b]\} = F\{[a,b]\}$. We write $F_n \xrightarrow{D} F$, when F_n, F are probability distribution functions (equivalent to $\lim_{n\to\infty} F_n(a) = F(a)$ for all continuity points a of F).

For $F \in \mathcal{M}(\mathbb{R})$,

$$m_F(z) \equiv \int \frac{1}{x-z} dF(x), \quad z \in \mathbb{C}^+ \equiv \{z \in \mathbb{C} : \Im z > 0\}$$

is defined as the Stieltjes transform of F.

Below are some fundamental properties of Stieltjes transforms:

(1) m_F is an analytic function on $\mathbb{C}^+$.
(2) $\Im m_F(z) > 0$.
(3) $|m_F(z)| \leq \frac{1}{\Im z}$.

(4) For continuity points $a < b$ of F

$$F\{[a,b]\} = \frac{1}{\pi} \lim_{\eta \to 0^+} \int_a^b \Im m_F(\xi + i\eta) d\xi,$$

since the right hand side

$$= \frac{1}{\pi} \lim_{\eta \to 0^+} \int_a^b \int \frac{\eta}{(x-\xi)^2 + \eta^2} dF(x) d\xi$$

$$= \frac{1}{\pi} \lim_{\eta \to 0^+} \int \int_a^b \frac{\eta}{(x-\xi)^2 + \eta^2} d\xi dF(x)$$

$$= \frac{1}{\pi} \lim_{\eta \to 0^+} \int \left[\mathrm{Tan}^{-1}\left(\frac{b-x}{\eta}\right) - \mathrm{Tan}^{-1}\left(\frac{a-x}{\eta}\right) \right] dF(x)$$

$$= \int I_{[a,b]} dF(x) = F\{[a,b]\}.$$

(5) If, for $x_0 \in \mathbb{R}$, $\Im m_F(x_0) \equiv \lim_{z \in \mathbb{C}^+ \to x_0} \Im m_F(z)$ exists, then F is differentiable at x_0 with value $(\frac{1}{\pi}) \Im m_F(x_0)$ ([9]).

Let $S \subset \mathbb{C}^+$ be countable with a cluster point in $\mathbb{C}^+$. Using ([4]), the fact that $F_n \xrightarrow{v} F$ is equivalent to

$$\int f_n(x) dF_n(x) \to \int f(x) dF(x)$$

for all continuous f vanishing at $\pm\infty$, and the fact that an analytic function defined on $\mathbb{C}^+$ is uniquely determined by the values it takes on S, we have

$$F_n \xrightarrow{v} F \iff m_{F_n}(z) \to m_F(z) \quad \text{for all } z \in S.$$

The fundamental connection to random matrices is:

For any Hermitian $n \times n$ matrix A, we let F^A denote the *empirical distribution function* (e.d.f.) of its eigenvalues:

$$F^A(x) = \frac{1}{n}(\text{number of eigenvalues of } A \leq x).$$

Then

$$m_{F^A}(z) = \frac{1}{n} \mathrm{tr}\,(A - zI)^{-1}.$$

So, if we have a sequence $\{A_n\}$ of Hermitian random matrices, to show, with probability one, $F^{A_n} \xrightarrow{v} F$ for some $F \in \mathcal{M}(\mathbb{R})$, it is equivalent to show for any $z \in \mathbb{C}^+$

$$\frac{1}{n}\operatorname{tr}(A_n - zI)^{-1} \to m_F(z) \quad \text{a.s.}$$

The main goal of the lectures is to show the importance of the Stieltjes transform to limiting behavior of certain classes of random matrices. We will begin with an attempt at providing a systematic way to show a.s. convergence of the e.d.f.'s of the eigenvalues of three classes of large dimensional random matrices via the Stieltjes transform approach. Essential properties involved will be emphasized in order to better understand where randomness comes in and where basic properties of matrices are used.

Then it will be shown, via the Stieltjes transform, how the limiting distribution can be numerically constructed, how it can explicitly (mathematically) be derived in some cases, and, in general, how important qualitative information can be inferred. Other results will be reviewed, namely the exact separation properties of eigenvalues, and distributional behavior of linear spectral statistics.

It is hoped that with this knowledge other ensembles can be explored for possible limiting behavior.

Each theorem below corresponds to a matrix ensemble. For each one the random quantities are defined on a common probability space. They all assume:

For $n = 1, 2, \ldots$ $X_n = (X_{ij}^n)$, $n \times N$, $X_{ij}^n \in \mathbb{C}$, i.d. for all n, i, j, independent across i, j for each n, $\mathsf{E}|X_{1\,1}^1 - \mathsf{E}X_{1\,1}^1|^2 = 1$, and $N = N(n)$ with $n/N \to c > 0$ as $n \to \infty$.

Theorem 1.1. ([6], [8]). *Assume:*

(a) $T_n = \operatorname{diag}(t_1^n, \ldots, t_n^n)$, $t_i^n \in \mathbb{R}$, *and the e.d.f. of* $\{t_1^n, \ldots, t_n^n\}$ *converges weakly, with probability one, to a nonrandom probability distribution function* H *as* $n \to \infty$.
(b) A_n *is a random* $N \times N$ *Hermitian random matrix for which* $F^{A_n} \xrightarrow{v} \mathcal{A}$ *where* $\mathcal{A}$ *is nonrandom (possibly defective).*
(c) X_n, T_n, *and* A_n *are independent.*

Let $B_n = A_n + (1/N)X_n^* T_n X_n$. *Then, with probability one,* $F^{B_n} \xrightarrow{v} \hat{F}$ *as*

$n \to \infty$ where for each $z \in \mathbb{C}^+$ $m = m_{\hat{F}}(z)$ satisfies

$$m = m_{\mathcal{A}}\left(z - c\int \frac{t}{1+tm}dH(t)\right). \tag{1.1}$$

It is the only solution to (1.1) with positive imaginary part.

Theorem 1.2. ([10], [7]). *Assume:*
T_n $n \times n$ is random Hermitian non-negative definite, independent of X_n with $F^{T_n} \xrightarrow{D} H$ a.s. as $n \to \infty$, H nonrandom.

Let $T_n^{1/2}$ denote any Hermitian square root of T_n, and define $B_n = (1/N)T_n^{1/2}XX^*T_n^{1/2}$. Then, with probability one, $F^{B_n} \xrightarrow{D} F$ as $n \to \infty$ where for each $z \in \mathbb{C}^+$ $m = m_F(z)$ satisfies

$$m = \int \frac{1}{t(1-c-czm)-z}dH(t). \tag{1.2}$$

It is the only solution to (1.2) in the set $\{m \in \mathbb{C} : -(1-c)/z + cm \in \mathbb{C}^+\}$.

Theorem 1.3. ([3]). *Assume:*
R_n $n \times N$ is random, independent of X_n, with $F^{(1/N)R_nR_n^*} \xrightarrow{D} H$ a.s. as $n \to \infty$, H nonrandom.

Let $B_n = (1/N)(R_n + \sigma X_n)(R_n + \sigma X_n)^*$ where $\sigma > 0$, nonrandom. Then, with probability one, $F^{B_n} \xrightarrow{D} F$ as $n \to \infty$ where for each $z \in \mathbb{C}^+$ $m = m_F(z)$ satisfies

$$m = \int \frac{1}{\frac{t}{1+\sigma^2 cm} - (1+\sigma^2 cm)z + \sigma^2(1-c)}dH(t). \tag{1.3}$$

It is the only solution to (1.3) in the set $\{m \in \mathbb{C}^+ : \Im(mz) \geq 0\}$.

Remark 1.4. In Theorem 1.1, if $A_n = 0$ for all n large, then $m_{\mathcal{A}}(z) = -1/z$ and we find that m_F has an inverse

$$z = -\frac{1}{m} + c\int \frac{t}{1+tm}dH(t). \tag{1.4}$$

Since

$$F^{(1/N)X_n^*T_nX_n} = \left(1 - \frac{n}{N}\right)I_{[0,\infty)} + \frac{n}{N}F^{(1/N)T_n^{1/2}X_nX_n^*T_n^{1/2}}$$

we have

$$m_{F^{(1/N)X_n^*T_nX_n}}(z) = -\frac{1-n/N}{z} + \frac{n}{N}m_{F^{(1/N)T_n^{1/2}X_nX_n^*T_n^{1/2}}}(z) \quad z \in \mathbb{C}^+, \tag{1.5}$$

so we have

$$m_{\hat{F}}(z) = -\frac{1-c}{z} + cm_F(z). \tag{1.6}$$

Using this identity, it is easy to see that (1.2) and (1.4) are equivalent.

2. Why These Theorems are True

We begin with three facts which account for most of why the limiting results are true, and the appearance of the limiting equations for the Stieltjes transforms.

Lemma 2.1. *For $n \times n$ A, $q \in \mathbb{C}^n$, and $t \in \mathbb{C}$ with A and $A + tqq^*$ invertible, we have*

$$q^*(A + tqq^*)^{-1} = \frac{1}{1 + tq^*A^{-1}q} q^*A^{-1}$$

(since $q^*A^{-1}(A + tqq^*) = (1 + tq^*A^{-1}q)q^*$).

Corollary 2.1. *For $q = a + b$, $t = 1$ we have*

$$a^*(A + (a+b)(a+b)^*)^{-1} = a^*A^{-1} - \frac{a^*A^{-1}(a+b)}{1 + (a+b)^*A^{-1}(a+b)}(a+b)^*A^{-1}$$

$$= \frac{1 + b^*A^{-1}(a+b)}{1 + (a+b)^*A^{-1}(a+b)} a^*A^{-1} - \frac{a^*A^{-1}(a+b)}{1 + (a+b)^*A^{-1}(a+b)} b^*A^{-1}.$$

Proof. Using Lemma 2.1, we have

$$(A + (a+b)(a+b)^*)^{-1} - A^{-1} = -(A + (a+b)(a+b)^*)^{-1}(a+b)(a+b)^*A^{-1}$$

$$= -\frac{1}{1 + (a+b)^*A^{-1}(a+b)} A^{-1}(a+b)(a+b)^*A^{-1}.$$

Multiplying both sides on the left by a^* gives the result. □

Lemma 2.2. *For $n \times n$ A and B, with B Hermitian, $z \in \mathbb{C}^+$, $t \in \mathbb{R}$, and $q \in \mathbb{C}^n$, we have*

$$|\mathrm{tr}\,[(B - zI)^{-1} - (B + tqq^* - zI)^{-1}]A| = \left| t\frac{q^*(B - zI)^{-1}A((B - zI)^{-1}q}{1 + tq^*(B - zI)^{-1}q} \right| \le \frac{\|A\|}{\Im z}.$$

Proof. The identity follows from Lemma 2.1. We have

$$\left|t\frac{q^*(B-zI)^{-1}A((B-zI)^{-1}q}{1+tq^*(B-zI)^{-1}q}\right| \le \|A\||t|\,\frac{\|(B-zI)^{-1}q\|^2}{|1+tq^*(B-zI)^{-1}q|}.$$

Write $B=\sum_i \lambda_i e_i e_i^*$, its spectral decomposition. Then

$$\|(B-zI)^{-1}q\|^2 = \sum_i \frac{|e_i^*q|^2}{|\lambda_i - z|^2}$$

and

$$|1+tq^*(B-zI)^{-1}q| \ge |t|\Im(q^*(B-zI)^{-1}q) = |t|\Im z\sum_i \frac{|e_i^*q|^2}{|\lambda_i - z|^2}. \qquad \square$$

Lemma 2.3. *For $X=(X_1,\dots,X_n)^T$ i.i.d. standardized entries, C $n\times n$, we have for any $p\ge 2$*

$$\mathsf{E}|X^*CX - \operatorname{tr} C|^p \le K_p\big((\mathsf{E}|X_1|^4 \operatorname{tr} CC^*)^{p/2} + \mathsf{E}|X_1|^{2p}\operatorname{tr}(CC^*)^{p/2}\big)$$

where the constant K_p does not depend on n, C, nor on the distribution of X_1. (Proof given in [1].)

From these properties, roughly speaking, we can make observations like the following: for $n\times n$ Hermitian A, $q=(1/\sqrt{n})(X_1,\dots,X_n)^T$, with X_i i.i.d. standardized and independent of A, and $z\in\mathbb{C}^+$, $t\in\mathbb{R}$

$$tq^*(A+tqq^*-zI)^{-1}q = \frac{tq^*(A-zI)^{-1}q}{1+tq^*(A-zI)^{-1}q}$$

$$= 1-\frac{1}{1+tq^*(A-zI)^{-1}q} \approx 1-\frac{1}{1+t(1/n)\operatorname{tr}(A-zI)^{-1}}$$

$$\approx 1-\frac{1}{1+t\,m_{A+tqq^*}(z)}.$$

Making this and other observations rigorous requires technical considerations, the first being truncation and centralization of the elements of X_n, and truncation of the eigenvalues of T_n in Theorem 1.2 (not needed in Theorem 1.1) and $(1/n)R_nR_n^*$ in Theorem 1.3, all at a rate slower than n ($a\ln n$ for some positive a is sufficient). The truncation and centralization steps will be outlined later. We are at this stage able to go through algebraic manipulations, keeping in mind the above three lemmas, and intuitively derive the equations appearing in each of the three theorems. At the same time we can see what technical details need to be worked out.

Before continuing, two more basic properties of matrices are included here.

Lemma 2.4. *Let* $z_1, z_2 \in \mathbb{C}^+$ *with* $\max(\Im z_1, \Im z_2) \geq v > 0$, A *and* B $n \times n$ *with* A *Hermitian, and* $q \in \mathbb{C}C^n$. *Then*

$$|\mathrm{tr}\, B((A - z_1 I)^{-1} - (A - z_2 I)^{-1})| \leq |z_2 - z_1| N \|B\| \frac{1}{v^2}, \text{ and}$$

$$|q^* B(A - z_1 I)^{-1} q - q^* B(A - z_2 I)^{-1} q| \leq |z_2 - z_1| \, \|q\|^2 \|B\| \frac{1}{v^2}.$$

Consider first the B_n in Theorem 1.1. Let q_i denote $1/\sqrt{N}$ times the ith column of X_n^*. Then

$$(1/N) X_n^* T_n X_n = \sum_{i=1}^{n} t_i q_i q_i^*.$$

Let $B_{(i)} = B_n - t_i q_i q_i^*$. For any $z \in \mathbb{C}^+$ and $x \in \mathbb{C}$ we write

$$B_n - zI = A_n - (z - x)I + (1/N) X_n^* T_n X_n - xI.$$

Taking inverses we have

$$(A_n - (z - x)I)^{-1}$$

$$= (B_n - zI)^{-1} + (A_n - (z - x)I)^{-1}((1/N) X_n^* T_n X_n - xI)(B_n - zI)^{-1}.$$

Dividing by N, taking traces and using Lemma 2.1 we find

$$m_{F^{A_n}}(z - x) - m_{F^{B_n}}(z)$$

$$= (1/N) \mathrm{tr}\, (A_n - (z - x)I)^{-1} \left(\sum_{i=1}^{n} t_i q_i q_i^* - xI \right) (B_n - zI)^{-1}$$

$$= (1/N) \sum_{i=1}^{n} \frac{t_i q_i^* (B_{(i)} - zI)^{-1} (A_n - (z - x)I)^{-1} q_i}{1 + t_i q_i^* (B_{(i)} - zI)^{-1} q_i}$$

$$- x (1/N) \mathrm{tr}\, (B_n - zI)^{-1} (A_n - (z - x)I)^{-1}.$$

Notice when x and q_i are independent, Lemmas 2.2, 2.3 give us

$$q_i^* (B_{(i)} - zI)^{-1} (A_n - (z - x)I)^{-1} q_i \approx (1/N) \mathrm{tr}\, (B_n - zI)^{-1} (A_n - (z - x)I)^{-1}.$$

Letting

$$x = x_n = (1/N) \sum_{i=1}^{n} \frac{t_i}{1 + t_i m_{F^{B_n}}(z)}$$

we have

$$m_{F^{A_n}}(z-x_n) - m_{F^{B_n}}(z) = (1/N)\sum_{i=1}^{n} \frac{t_i}{1+t_i m_{F^{B_n}}(z)} d_i \tag{2.1}$$

where

$$d_i = \frac{1+t_i m_{F^{B_n}}(z)}{1+t_i q_i^*(B_{(i)}-zI)^{-1}q_i} q_i^*(B_{(i)}-zI)^{-1}(A_n-(z-x_n)I)^{-1}q_i$$

$$-(1/N)\mathsf{tr}\,(B_n-zI)^{-1}(A_n-(z-x_n)I)^{-1}\,.$$

In order to use Lemma 2.3, for each i, x_n is replaced by

$$x_{(i)} = (1/N)\sum_{j=1}^{n} \frac{t_j}{1+t_j m_{F^{B_{(i)}}}(z)}.$$

An outline of the remainder of the proof is given. It is easy to argue that if $\mathcal{A}$ is the zero measure on $\mathbb{R}$ (that is, almost surely, only $o(N)$ eigenvalues of A_n remain bounded), then the Stieltjes transforms of F^{A_n} and F^{B_n} converge a.s. to zero, the limits obviously satisfying (1.1). So we assume $\mathcal{A}$ is not the zero measure. One can then show

$$\delta = \inf_n \Im(m_{F^{B_n}}(z))$$

is positive almost surely.

Using Lemma 2.3 ($p=6$ is sufficient) and the fact that all matrix inverses encountered are bounded in spectral norm by $1/\Im z$ we have from standard arguments using Boole's and Chebyshev's inequalities, almost surely

$$\max_{i\le n} \max[|\,\|q_i\|^2-1|, |q_i^*(B_{(i)}-zI)^{-1}q_i - m_{F^{B_{(i)}}}(z)|, \tag{2.2}$$

$$|q_i^*(B_{(i)}-zI)^{-1}(A_n-(z-x_{(i)})I)^{-1}q_i - \frac{1}{N}\mathsf{tr}\,(B_{(i)}-zI)^{-1}(A_n-(z-x_{(i)})I)^{-1}|]$$

$$\to 0 \quad \text{as } n\to\infty.$$

Consider now a realization for which (2.2) holds, $\delta>0$, $F^{T_n} \xrightarrow{D} H$, and $F^{A_n} \xrightarrow{v} \mathcal{A}$. From Lemma 2.2 and (2.2) we have

$$\max_{i\le n} \max[|m_{F^{B_n}}(z) - m_{F^{B_{(i)}}}(z)|, |m_{F^{B_n}}(z) - q_i^*(B_{(i)}-zI)^{-1}q_i|] \to 0, \tag{2.3}$$

and subsequently

$$\max_{i\le n} \max\left[\left|\frac{1+t_i m_{F^{B_n}}(z)}{1+t_i q_i^*(B_{(i)}-zI)^{-1}q_i} - 1\right|, |x - x_{(i)}|\right] \to 0. \tag{2.4}$$

Therefore, from Lemmas 2.2, 2.4, and (2.2)-(2.4), we get $\max_{i\leq n} d_i \to 0$, and since

$$\left|\frac{t_i}{1+t_i m_{F^{B_n}}(z)}\right| \leq \frac{1}{\delta},$$

we conclude from (2.1) that

$$m_{F^{A_n}}(z-x_n) - m_{F^{B_n}}(z) \to 0.$$

Consider a subsequence $\{n_i\}$ on which $m_{F^{B_{n_i}}}(z)$ converges to a number m. It follows that

$$x_{n_i} \to c\int \frac{t}{1+tm} dH(t).$$

Therefore, m satisfies (1.1). Uniqueness (to be discussed later) gives us, for this realization $m_{F^{B_n}}(z) \to m$. This event occurs with probability one.

3. The Other Equations

Let us now derive the equation for the matrix $B_n = (1/N)T_n^{1/2}X_nX_n^*T_n^{1/2}$, after the truncation steps have been taken. Let $c_n = n/N$, $q_j = (1/\sqrt{n})X_{\cdot j}$ (the jth column of X_n), $r_j = (1/\sqrt{N})T_n^{1/2}X_{\cdot j}$, and $B_{(j)} = B_n - r_jr_j^*$. Fix $z \in \mathbb{C}^+$ and let $m_n(z) = m_{F^{B_n}}(z)$, $\mathbf{m}_n(z) = m_{F^{(1/N)X_n^*T_nX_n}}(z)$. By (1.5) we have

$$\mathbf{m}_n(z) = -\frac{1-c_n}{z} + c_n m_n. \tag{3.1}$$

We first derive an identity for $\mathbf{m}_n(z)$. Write

$$B_n - zI + zI = \sum_{j=1}^{N} r_jr_j^*.$$

Taking the inverse of $B_n - zI$ on the right on both sides and using Lemma 2.1, we find

$$I + z(B_n - zI)^{-1} = \sum_{j=1}^{N} \frac{1}{1+r_j^*(B_{(j)}-zI)^{-1}r_j} r_jr_j^*(B_{(j)}-zI)^{-1}.$$

Taking the trace on both sides and dividing by N we have

$$\begin{aligned} c_n + zc_nm_n &= \frac{1}{N}\sum_{j=1}^{N} \frac{r_j^*(B_{(j)}-zI)^{-1}r_j}{1+r_j^*(B_{(j)}-zI)^{-1}r_j} \\ &= 1 - \frac{1}{N}\sum_{j=1}^{N} \frac{1}{1+r_j^*(B_{(j)}-zI)^{-1}r_j}. \end{aligned}$$

Therefore

$$\mathbf{m}_n(z) = -\frac{1}{N}\sum_{j=1}^{N} \frac{1}{z(1+r_j^*(B_{(j)}-zI)^{-1}r_j)}. \tag{3.2}$$

Write $B_n - zI - (-z\mathbf{m}_n(z)T_n - zI) = \sum_{j=1}^N r_j r_j^* - (-z\mathbf{m}_n(z))T_n$. Taking inverses and using Lemma 2.1, (3.2) we have

$$(-z\mathbf{m}_n(z)T_n - zI)^{-1} - (B_n - zI)^{-1}$$

$$= (-z\mathbf{m}_n(z)T_n - zI)^{-1}\left[\sum_{j=1}^{N} r_j r_j^* - (-z\mathbf{m}_n(z))T_n\right](B_n - zI)^{-1}$$

$$= \sum_{j=1}^{N} \frac{-1}{z(1+r_j^*(B_{(j)}-zI)^{-1}r_j)}\big[(\mathbf{m}_n(z)T_n+I)^{-1}r_j r_j^*(B_{(j)}-zI)^{-1}$$

$$- (1/N)(\mathbf{m}_n(z)T_n+I)^{-1}T_n(B_n-zI)^{-1}\big].$$

Taking the trace and dividing by n we find

$$(1/n)\mathsf{tr}\,(-z\mathbf{m}_n(z)T_n - zI)^{-1} - m_n(z) = \frac{1}{N}\sum_{j=1}^{N}\frac{-1}{z(1+r_j^*(B_{(j)}-zI)^{-1}r_j)}d_j$$

where

$$d_j = q_j^* T_n^{1/2}(B_{(j)}-zI)^{-1}(\mathbf{m}_n(z)T_n+I)^{-1}T_n^{1/2}q_j$$

$$- (1/n)\mathsf{tr}\,(\mathbf{m}_n(z)T_n+I)^{-1}T_n(B_n-zI)^{-1}.$$

The derivation for Theorem 1.3 will proceed in a constructive way. Here we let x_j and r_j denote, respectively, the jth columns of X_n and R_n (after truncation). As before $m_n = m_{F^{B_n}}$, and let

$$\mathbf{m}_n(z) = m_{F^{(1/N)(R_n+\sigma X_n)^*(R_n+\sigma X_n)}}(z).$$

We have again the relationship (3.1). Notice then equation (1.3) can be written

$$m = \int \frac{1}{\frac{t}{1+\sigma^2 cm} - \sigma^2 z\mathbf{m} - z}dH(t) \tag{3.3}$$

where

$$\mathbf{m} = -\frac{1-c}{z} + cm.$$

Let $B_{(j)} = B_n - (1/N)(r_j+\sigma x_j)(r_j+\sigma x_j)^*$. Then, as in (3.2) we have

$$\mathbf{m}_n(z) = -\frac{1}{N}\sum_{j=1}^{N}\frac{1}{z(1+(1/N)(r_j+\sigma x_j)^*(B_{(j)}-zI)^{-1}(r_j+\sigma x_j))}. \tag{3.4}$$

Pick $z \in \mathbb{C}^+$. For any $n \times n$ Y_n we write

$$B_n - zI - (Y_n - zI) = \frac{1}{N}\sum_{j=1}^{N}(r_j + \sigma x_j)(r_j + \sigma x_j)^* - Y_n.$$

Taking inverses, dividing by n and using Lemma 2.1 we get

$$(1/n)\mathrm{tr}\,(Y_n - zI)^{-1} - m_n(z)$$

$$= \frac{1}{N}\sum_{j=1}^{N}\frac{(1/n)(r_j + \sigma x_j)^*(B_{(j)} - zI)^{-1}(Y_n - zI)^{-1}(r_j + \sigma x_j)}{1 + (1/N)(r_j + \sigma x_j)^*(B_{(j)} - zI)^{-1}(r_j + \sigma x_j)}$$

$$- (1/n)\mathrm{tr}\,(Y_n - zI)^{-1}Y_n(B_n - zI)^{-1}.$$

The goal is to determine Y_n so that each term goes to zero. Notice first that

$$(1/n)x_j^*(B_{(j)} - zI)^{-1}(Y_n - zI)^{-1}x_j \approx (1/n)\mathrm{tr}\,(B_n - zI)^{-1}(Y_n - zI)^{-1},$$

so from (3.4) we see that Y_n should have a term

$$-\sigma^2 z\mathbf{m}_n(z)I.$$

Since for any $n \times n$ C bounded in norm

$$|(1/n)x_j^* C r_j|^2 = (1/n^2)x_j^* C r_j r_j^* C^* x_j$$

we have from Lemma 2.3

$$|(1/n)x_j^* C r_j|^2 \approx (1/n^2)\mathrm{tr}\, C r_j r_j^* C^* = (1/n^2) r_j^* C^* C r_j = o(1) \qquad (3.5)$$

(from truncation $(1/N)\|r_j\|^2 \leq \ln n$), so the cross terms are negligible.

This leaves us $(1/n)r_j^*(B_{(j)} - zI)^{-1}(Y_n - zI)^{-1}r_j$. Recall Corollary 2.1:

$$a^*(A + (a+b)(a+b)^*)^{-1}$$

$$= \frac{1 + b^* A^{-1}(a+b)}{1 + (a+b)^* A^{-1}(a+b)} a^* A^{-1} - \frac{a^* A^{-1}(a+b)}{1 + (a+b)^* A^{-1}(a+b)} b^* A^{-1}.$$

Identify a with $(1/\sqrt{N})r_j$, b with $(1/\sqrt{N})\sigma x_j$, and A with $B_{(j)}$. Using Lemmas 2.2, 2.3 and (3.5), we have

$$(1/n)r_j^*(B_n - zI)^{-1}(Y_n - zI)^{-1}r_j$$

$$\approx \frac{1 + \sigma^2 c_n m_n(z)}{1 + \frac{1}{N}(r_j + \sigma x_j)^*(B_{(j)} - zI)^{-1}(r_j + \sigma x_j)} \frac{1}{n} r_j^*(B_{(j)} - zI)^{-1}(Y_n - zI)^{-1}r_j.$$

Therefore

$$\frac{1}{N}\sum_{j=1}^{N}\frac{(1/n)r_j^*(B_{(j)}-zI)^{-1}(Y_n-zI)^{-1}r_j}{1+\frac{1}{N}(r_j+\sigma x_j)^*(B_{(j)}-zI)^{-1}(r_j+\sigma x_j)}$$

$$\approx\frac{1}{N}\sum_{j=1}^{N}\frac{(1/n)r_j^*(B_n-zI)^{-1}(Y_n-zI)^{-1}r_j}{1+\sigma^2c_nm_n(z)}$$

$$=(1/n)\frac{1}{1+\sigma^2c_nm_n(z)}\mathrm{tr}\,(1/N)R_nR_n^*(B_n-zI)^{-1}(Y_n-zI)^{-1}.$$

So we should take

$$Y_n=\frac{1}{1+\sigma^2c_nm_n(z)}(1/N)R_nR_n^*-\sigma^2z\mathbf{m}_n(z)I.$$

Then $(1/n)\mathrm{tr}\,(Y_n-zI)^{-1}$ will approach the right hand side of (3.3).

4. Proof of Uniqueness of (1.1)

For $m\in\mathbb{C}^+$ satisfying (1.1) with $z\in\mathbb{C}^+$ we have

$$m=\int\frac{1}{\tau-\left(z-c\int\frac{t}{1+tm}dH(t)\right)}d\mathcal{A}(\tau)$$

$$=\int\frac{1}{\tau-\Re\left(z-c\int\frac{t}{1+tm}dH(t)\right)-i\left(\Im z+c\int\frac{t^2\Im m}{|1+tm|^2}dH(t)\right)}d\mathcal{A}(\tau).$$

Therefore

$$\Im m=\left(\Im z+c\int\frac{t^2\Im m}{|1+tm|^2}dH(t)\right)\int\frac{1}{\left|\tau-z+c\int\frac{t}{1+tm}dH(t)\right|^2}d\mathcal{A}(\tau). \tag{4.1}$$

Suppose $\mathbf{m}\in\mathbb{C}^+$ also satisfies (1.1). Then

$$m-\mathbf{m}=c\int\frac{\left[\int\frac{t}{1+t\mathbf{m}}-\frac{t}{1+tm}\right]dH(t)}{\left(\tau-z+c\int\frac{t}{1+tm}dH(t)\right)\left(\tau-z+c\int\frac{t}{1+t\mathbf{m}}dH(t)\right)}d\mathcal{A}(\tau)$$

$$\times(m-\mathbf{m})c\int\frac{t^2}{(1+tm)(1+t\mathbf{m})}dH(t) \tag{4.2}$$

$$\times \int \frac{1}{\left(\tau - z + c\int \frac{t}{1+tm} dH(t)\right)\left(\tau - z + c\int \frac{t}{1+t\mathbf{m}} dH(t)\right)} d\mathcal{A}(\tau).$$

Using Cauchy-Schwarz and (4.1) we have

$$\left| c\int \frac{t^2}{(1+tm)(1+t\mathbf{m})} dH(t) \right.$$

$$\left. \times \int \frac{1}{\left(\tau - z + c\int \frac{t}{1+tm} dH(t)\right)\left(\tau - z + c\int \frac{t}{1+t\mathbf{m}} dH(t)\right)} d\mathcal{A}(\tau) \right|$$

$$\leq \left(c\int \frac{t^2}{|1+tm|^2} dH(t) \int \frac{1}{\left|\tau - z + c\int \frac{t}{1+tm} dH(t)\right|^2} d\mathcal{A}(\tau) \right)^{1/2}$$

$$\times \left(c\int \frac{t^2}{|1+t\mathbf{m}|^2} dH(t) \int \frac{1}{\left|\tau - z + c\int \frac{t}{1+t\mathbf{m}} dH(t)\right|^2} d\mathcal{A}(\tau) \right)^{1/2}$$

$$= \left(c\int \frac{t^2}{|1+tm|^2} dH(t) \frac{\Im m}{\left(\Im z + c\int \frac{t^2 \Im m}{|1+tm|^2} dH(t)\right)} \right)^{1/2}$$

$$\times \left(c\int \frac{t^2}{|1+t\mathbf{m}|^2} dH(t) \frac{\Im \mathbf{m}}{\left(\Im z + c\int \frac{t^2 \Im \mathbf{m}}{|1+t\mathbf{m}|^2} dH(t)\right)} \right)^{1/2} < 1.$$

Therefore, from (4.2) we must have $m = \mathbf{m}$.

5. Truncation and Centralization

We outline here the steps taken to enable us to assume in the proof of Theorem 1.1, for each n, the X_{ij}'s are bounded by a multiple of $\ln n$. The following lemmas are needed.

Lemma 5.1. *Let $X_1, \ldots, X_n$ be i.i.d. Bernoulli with $p = \mathsf{P}(X_1 = 1) < 1/2$. Then for any $\epsilon > 0$ such that $p + \epsilon \leq 1/2$ we have*

$$\mathsf{P}\left(\frac{1}{n}\sum_{i=1}^{n} X_i - p \geq \epsilon\right) \leq e^{-\frac{n\epsilon^2}{2(p+\epsilon)}}.$$

Lemma 5.2. *Let A be $N \times N$ Hermitian, Q, $\overline{Q}$ both $n \times N$, and T, $\overline{T}$ both $n \times n$ Hermitian. Then*

(a) $$\|F^{A+Q^*TQ} - F^{A+\overline{Q}^*T\overline{Q}}\| \le \frac{2}{N}\text{rank}(Q - \overline{Q})$$

and

(b) $$\|F^{A+Q^*TQ} - F^{A+Q^*\overline{T}Q}\| \le \frac{1}{N}\text{rank}(T - \overline{T}).$$

Lemma 5.3. *For rectangular A, $\text{rank}(A) \le$ the number of nonzero entries of A.*

Lemma 5.4. *For Hermitian $N \times N$ matrices A, B*

$$\sum_{i=1}^{N}(\lambda_i^A - \lambda_i^B)^2 \le \mathsf{tr}\,(A-B)^2.$$

Lemma 5.5. *Let $\{f_i\}$ be an enumeration of all continuous functions that take a constant $\frac{1}{m}$ value (m a positive integer) on $[a,b]$, where a, b are rational, 0 on $(-\infty, a - \frac{1}{m}] \cup [b + \frac{1}{m}, \infty)$, and linear on each of $[a - \frac{1}{m}, a]$, $[b, b + \frac{1}{m}]$. Then*

(a) *for $F_1, F_2 \in \mathcal{M}(\mathbb{R})$*

$$D(F_1, F_2) \equiv \sum_{i=1}^{\infty}\left|\int f_i dF_1 - \int f_i dF_2\right| 2^{-i}$$

is a metric on $\mathcal{M}(\mathbb{R})$ inducing the topology of vague convergence.

(b) *For $F_N, G_N \in \mathcal{M}(\mathbb{R})$*

$$\lim_{N\to\infty}\|F_N - G_N\| = 0 \implies \lim_{N\to\infty} D(F_N, G_N) = 0.$$

(c) *For empirical distribution functions F, G on the (respective) sets $\{x_1, \ldots, x_N\}, \{y_1, \ldots, y_N\}$*

$$D^2(F,G) \le \left(\frac{1}{N}\sum_{j=1}^{N}|x_j - y_j|\right)^2 \le \frac{1}{N}\sum_{j=1}^{N}(x_j - y_j)^2.$$

Let $p_n = \mathsf{P}(|X_{11}| \ge \sqrt{n})$. Since the second moment of X_{11} is finite we have

$$np_n = o(1). \tag{5.1}$$

Let $\widehat{X}_{ij} = X_{ij}I_{(|X_{ij}|<\sqrt{n})}$ and $\widehat{B}_n = A_n + (1/N)\widehat{X}_n^* T_n \widehat{X}_n$, where $\widehat{X} = (\widehat{X}_{ij})$. Then from Lemmas 5.2(a), 5.3, for any positive ϵ

$$\mathsf{P}(\|F^{B_n} - F^{\widehat{B}_n}\| \geq \epsilon) \leq \mathsf{P}\left(\frac{2}{N}\sum_{ij} I_{(|X_{ij}|\geq\sqrt{n})} \geq \epsilon\right)$$

$$= \mathsf{P}\left(\frac{1}{Nn}\sum_{ij} I_{(|X_{ij}|\geq\sqrt{n})} - p_n \geq \frac{\epsilon}{2n} - p_n\right).$$

Then by Lemma 5.1, for all n large

$$\mathsf{P}(\|F^{B_n} - F^{\widehat{B}_n}\| \geq \epsilon) \leq e^{-\frac{N\epsilon}{16}},$$

which is summable. Therefore

$$\|F^{B_n} - F^{\widehat{B}_n}\| \xrightarrow{a.s.} 0.$$

Let $\widetilde{B}_n = A_n + (1/N)\widetilde{X}_n T_n \widetilde{X}_n^*$ where $\widetilde{X}_n = \widehat{X}_n - \mathsf{E}\widehat{X}_n$. Since $\operatorname{rank}(\mathsf{E}\widehat{X}_n) \leq 1$, we have from Lemma 5.2(a)

$$\|F^{\widehat{B}_n} - F^{\widetilde{B}_n}\| \longrightarrow 0.$$

For $\alpha > 0$ define $T_\alpha = \operatorname{diag}(t_1^n I_{(|t_1^n|\leq\alpha)}, \ldots, t_n^n I_{(|t_n^n|\leq\alpha)})$, and let Q be any $n \times N$ matrix. If α and $-\alpha$ are continuity points of H, we have by Lemma 5.2(b)

$$\|F^{A_n+Q^*T_nQ} - F^{A_n+Q^*T_\alpha Q}\|$$

$$\leq \frac{1}{N}\operatorname{rank}(T_n - T_\alpha) = \frac{1}{N}\sum_{i=1}^{n} I_{(|t_i^n|>\alpha)} \xrightarrow{a.s.} cH\{[-\alpha,\alpha]^c\}.$$

It follows that if $\alpha = \alpha_n \to \infty$ then

$$\|F^{A_n+Q^*T_nQ} - F^{A_n+Q^*T_\alpha Q}\| \xrightarrow{a.s.} 0.$$

Let $\overline{X}_{ij} = \widetilde{X}_{ij}I_{(|X_{ij}|<\ln n)} - \mathsf{E}\widetilde{X}_{ij}I_{(|X_{ij}|<\ln n)}$, $\overline{X}_n = ((1/\sqrt{N})\overline{X}_{ij})$, $\overline{\overline{X}}_{ij} = \widetilde{X}_{ij} - \overline{X}_{ij}$, and $\overline{\overline{X}}_n = ((1/\sqrt{N})\overline{\overline{X}}_{ij})$. Then, from Lemmas 5.5(c) and 5.4 and simple applications of Cauchy-Schwarz we have

$$D^2(F^{A_n+\widetilde{X}_nT_\alpha\widetilde{X}_n^*}, F^{A+\overline{X}_nT_\alpha\overline{X}_n{}^*}) \leq \frac{1}{N}\mathsf{tr}\,(\widetilde{X}_nT_\alpha\widetilde{X}_n^* - \overline{X}_nT_\alpha\overline{X}_n{}^*)^2$$

$$\leq \frac{1}{N}[\mathsf{tr}\,(\overline{\overline{X}}_nT_\alpha\overline{\overline{X}}_n^*)^2 + 4\mathsf{tr}\,(\overline{\overline{X}}_nT_\alpha\overline{X}_n^*\overline{X}_nT_\alpha\overline{\overline{X}}_n^*)$$

$$+ 4(\mathsf{tr}\,(\overline{\overline{X}}_nT_\alpha\overline{X}_n^*\overline{X}_nT_\alpha\overline{\overline{X}}_n^*)\mathsf{tr}\,(\overline{\overline{X}}_nT_\alpha\overline{\overline{X}}_n^*)^2)^{1/2}].$$

We have

$$\operatorname{tr}(\overline{\overline{X}}_n T_\alpha \overline{\overline{X}}_n^*)^2 \leq \alpha^2 \operatorname{tr}(\overline{\overline{X}}\,\overline{\overline{X}}^*)^2$$

and

$$\operatorname{tr}(\overline{\overline{X}}_n T_\alpha \overline{X}_n^* \overline{X} T_\alpha \overline{\overline{X}}^*) \leq (\alpha^4 \operatorname{tr}(\overline{\overline{X}}\,\overline{\overline{X}}^*)^2 \operatorname{tr}(\overline{X}\,\overline{X}^*)^2)^{1/2}.$$

Therefore, to verify

$$D(F^{A+\widetilde{X}T_\alpha \widetilde{X}^*}, F^{A+\overline{X}T_\alpha \overline{X}^*}) \xrightarrow{a.s.} 0$$

it is sufficient to find a sequence $\{\alpha_n\}$ increasing to ∞ so that

$$\alpha_n^4 \frac{1}{N} \operatorname{tr}(\overline{\overline{X}}\,\overline{\overline{X}}^*)^2 \xrightarrow{a.s.} 0 \text{ and } \frac{1}{N} \operatorname{tr}(\overline{X}\,\overline{X}^*)^2 = O(1) \text{ a.s.}$$

The details are omitted.

Notice the matrix $\operatorname{diag}(\mathsf{E}|\overline{X}_{11}|^2 t_1^n, \ldots, \mathsf{E}|\overline{X}_{11}|^2 t_n^n)$ also satisfies assumption (a) of Theorem 1.1. Just substitute this matrix for T_n, and replace $\overline{X}_n$ by $(1/\sqrt{\mathsf{E}|\overline{X}_{11}|^2})\overline{X}_n$. Therefore we may assume

(1) X_{ij} are i.i.d. for fixed n,
(2) $|X_{11}| \leq a \ln n$ for some positive a,
(3) $\mathsf{E}X_{11} = 0$, $\mathsf{E}|X_{11}|^2 = 1$.

6. The Limiting Distributions

The Stieltjes transform provides a great deal of information to the nature of the limiting distribution $\hat{F}$ when $A_n = 0$ in Theorem 1.1, and F in Theorems 1.2, 1.3. For the first two

$$z = -\frac{1}{\mathbf{m}} + c \int \frac{t}{1 + t\mathbf{m}} dH(t)$$

is the inverse of $\mathbf{m} = m_{\hat{F}}(z)$, the limiting Stieltjes transform of $F^{(1/N)X_n^* T_n X_n}$. Recall, when T_n is nonnegative definite, the relationships between F, the limit of $F^{(1/N)T_n^{1/2} X_n X_n^* T_n^{1/2}}$ and $\hat{F}$

$$\hat{F}(x) = 1 - cI_{[0,\infty)}(x) + cF(x),$$

and m_F and $m_{\hat{F}}$

$$m_{\hat{F}}(z) = -\frac{1-c}{z} + cm_F(z).$$

Based solely on the inverse of $m_{\hat{F}}$ the following is shown in [9]: (1) For all $x \in \mathbb{R}$, $x \neq 0$

$$\lim_{z \in \mathbb{C}^+ \to x} m_{\hat{F}}(z) \equiv m_0(x)$$

exists. The function m_0 is continuous on $\mathbb{R} - \{0\}$. Consequently, by property (5) of Stieltjes transforms, $\hat{F}$ has a continuous derivative f on $\mathbb{R} - \{0\}$ given by $\hat{f}(x) = \frac{1}{\pi}\Im m_0(x)$ (F subsequently has derivative $f = \frac{1}{c}\hat{f}$). The density $\hat{f}$ is analytic (possess a power series expansion) for every $x \neq 0$ for which $f(x) > 0$. Moreover, for these x, $\pi f(x)$ is the imaginary part of the unique $\mathbf{m} \in \mathbb{C}^+$ satisfying

$$x = -\frac{1}{\mathbf{m}} + c\int \frac{t}{1+t\mathbf{m}} dH(t).$$

(2) Let $x_{\hat{F}}$ denote the above function of $\mathbf{m}$. It is defined and analytic on $B \equiv \{\mathbf{m} \in \mathbb{R} : \mathbf{m} \neq 0, -\mathbf{m}^{-1} \in S_H^c\}$ (S_G^c denoting the complement of the support of distribution G). Then if $x \in S_{\hat{F}}^c$ we have $\mathbf{m} = m_0(x) \in B$ and $x'_{\hat{F}}(\mathbf{m}) > 0$. Conversely, if $\mathbf{m} \in B$ and $x = x'_{\hat{F}}(\mathbf{m}) > 0$, then $x \in S_{\hat{F}}^c$.

We see then a systematic way of determining the support of $\hat{F}$: Plot $x_{\hat{F}}(\mathbf{m})$ for $\mathbf{m} \in B$. Remove all intervals on the vertical axis corresponding to places where $x_{\hat{F}}$ is increasing. What remains is $S_{\hat{F}}$, the support of $\hat{F}$.

Let us look at an example where H places mass at 1, 3, and 10, with respective probabilities .2, .4, and .4, and $c = .1$. Figure (b) on the next page is the graph of

$$x_{\hat{F}}(\mathbf{m}) = -\frac{1}{\mathbf{m}} + .1\left(.2\frac{1}{1+\mathbf{m}} + .4\frac{3}{1+3\mathbf{m}} + .4\frac{10}{1+10\mathbf{m}}\right).$$

We see the support boundaries occur at relative extreme values. These values were estimated and for values of $x \in S_{\hat{F}}$, $f(x) = \frac{1}{c\pi}\Im m_0(x)$ was computed using Newton's method on $x = x_{\hat{F}}(\mathbf{m})$, resulting in figure (a).

It is possible for a support boundary to occur at a boundary of the support of B, which would only happen for a nondiscrete H. However, we have

(3) Suppose support boundary a is such that $m_{\hat{F}}(a) \in B$, and is a left-endpoint in the support of $\hat{F}$. Then for $x > a$ and near a

$$f(x) = \left(\int_a^x g(t)dt\right)^{1/2}$$

where $g(a) > 0$ (analogous statement holds for a a right-endpoint in the support of $\hat{F}$). Thus, near support boundaries, f and the square root function share common features, as can be seen in figure (a).

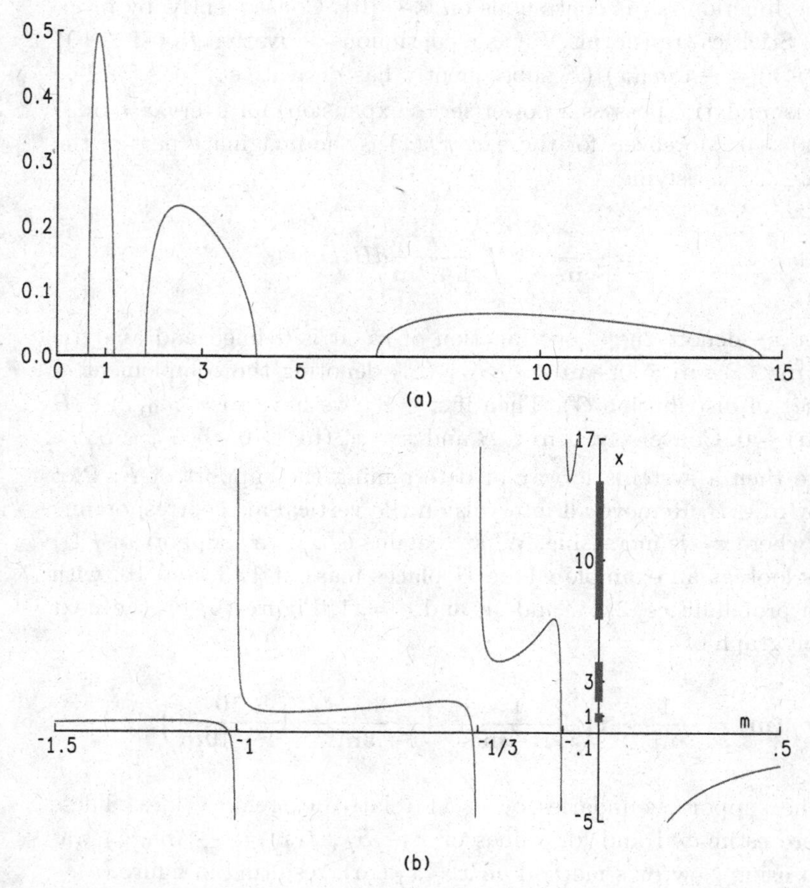

It is remarked here that similar results have been obtained for the matrices in Theorem 1.3. See [4].

Explicit solutions can be derived in a few cases. Consider the Marčenko-Pastur distribution, where $T_n = I$. Then $\mathbf{m} = m_0(x)$ solves

$$x = -\frac{1}{\mathbf{m}} + c\frac{1}{1+\mathbf{m}},$$

resulting in the quadratic equation

$$x\mathbf{m}^2 + \mathbf{m}(x + 1 - c) + 1 = 0$$

with solution

$$m = \frac{-(x+1-c) \pm \sqrt{(x+1-c)^2 - 4x}}{2x}$$

$$= \frac{-(x+1-c) \pm \sqrt{x^2 - 2x(1+c) + (1-c)^2}}{2x}$$

$$= \frac{-(x+1-c) \pm \sqrt{(x-(1-\sqrt{c})^2)(x-(1+\sqrt{c})^2)}}{2x}.$$

We see the imaginary part of m is zero when x lies outside the interval $[(1-\sqrt{c})^2, (1+\sqrt{c})^2]$, and we conclude that

$$f(x) = \begin{cases} \frac{\sqrt{(x-(1-\sqrt{c})^2)((1+\sqrt{c})^2-x)}}{2\pi c x} & x \in ((1-\sqrt{c})^2, (1+\sqrt{c})^2) \\ 0 & \text{otherwise.} \end{cases}$$

The Stieltjes transform in the multivariate F matrix case, that is, when $T_n = ((1/N')\underline{X}_n \underline{X}_n^*)^{-1}$, $\underline{X}_n$ $n \times N'$ containing i.i.d. standardized entries, $n/N' \to c' \in (0,1)$, also satisfies a quadratic equation. Indeed, H now is the distribution of the reciprocal of a Marčenko-Pastur distributed random variable which we'll denote by $X_{c'}$, the Stieltjes transform of its distribution denoted by $m_{X_{c'}}$. We have

$$x = -\frac{1}{\mathbf{m}} + c\mathsf{E}\left(\frac{\frac{1}{X_{c'}}}{1+\frac{1}{X_{c'}}\mathbf{m}}\right) = -\frac{1}{\mathbf{m}} + c\mathsf{E}\left(\frac{1}{X_{c'}+\mathbf{m}}\right)$$

$$= -\frac{1}{\mathbf{m}} + c m_{X_{c'}}(-\mathbf{m}).$$

From above we have

$$m_{X_{c'}}(z) = \frac{1-c'}{c'z} + \frac{-(z+1-c) + \sqrt{(z+1-c)^2 - 4z}}{2zc'}$$

$$= \frac{-z+1-c' + \sqrt{(z+1-c')^2 - 4z}}{2zc'}$$

(the square root defined so that the expression is a Stieltjes transform) so that $\mathbf{m} = m_0(x)$ satisfies

$$x = -\frac{1}{\mathbf{m}} + c\left(\frac{\mathbf{m}+1-c+\sqrt{(-\mathbf{m}+1-c)^2 + 4\mathbf{m}}}{-2\mathbf{m}c'}\right).$$

It follows that $\mathbf{m}$ satisfies

$$\mathbf{m}^2(c'x^2 + cx) + \mathbf{m}(2c'x - c^2 + c + cx(1-c')) + c' + c(1-c') = 0.$$

Solving for $\mathbf{m}$ we conclude that, with

$$b_1 = \left(\frac{1-\sqrt{1-(1-c)(1-c')}}{1-c'}\right)^2 \quad b_2 = \left(\frac{1-\sqrt{1-(1-c)(1-c')}}{1-c'}\right)^2$$

$$f(x) = \begin{cases} \frac{(1-c')\sqrt{(x-b_1)(b_2-x)}}{2\pi x(xc'+c)} & b_1 < x < b_2 \\ 0 & \text{otherwise.} \end{cases}$$

7. Other Uses of the Stieltjes Transform

We conclude these lectures with two results requiring Stieltjes transforms.

The first concerns the eigenvalues of matrices in Theorem 1.2 outside the support of the limiting distribution. The results mentioned so far clearly say nothing about the possibility of some eigenvalues lingering in this region. Consider this example with T_n given earlier, but now $c = .05$. Below is a scatterplot of the eigenvalues from a simulation with $n = 200$ ($N = 4000$), superimposed on the limiting density.

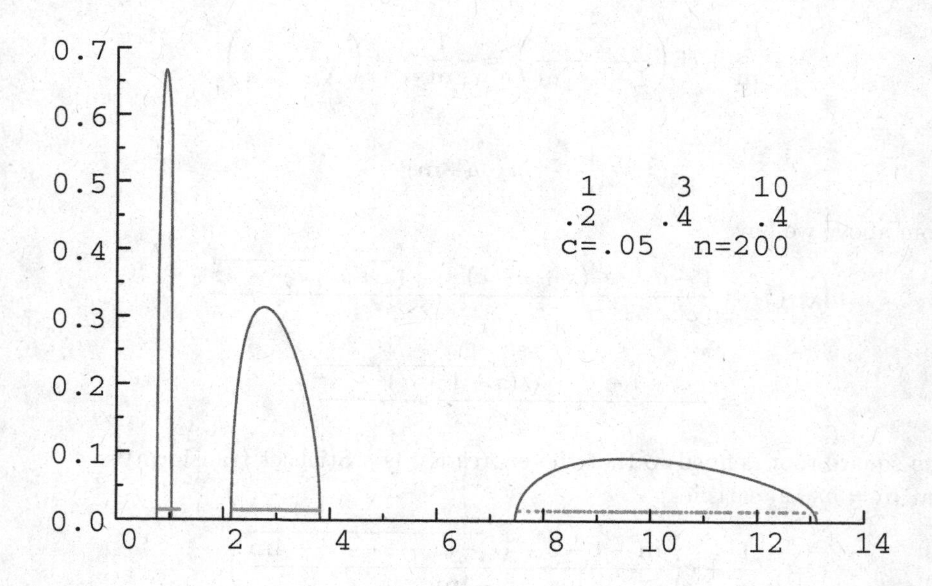

Here the entries of X_n are $N(0,1)$. All the eigenvalues appear to stay close to the limiting support. Such simulations were the prime motivation to prove

Theorem 7.1. ([1]). *Let, for any $d > 0$ and d.f. G, $\hat{F}^{d,G}$ denote the limiting e.d.f. of $(1/N)X_n^*T_nX_n$ corresponding to limiting ratio d and limiting $F^{T_n}G$.*

Assume in addition to the previous assumptions:

(a) $\mathsf{E}X_{11} = 0$, $\mathsf{E}|X_{11}|^2 = 1$, *and* $\mathsf{E}|X_{11}|^4 < \infty$.
(b) T_n *is nonrandom and* $\|T_n\|$ *is bounded in* n.
(c) *The interval* $[a, b]$ *with* $a > 0$ *lies in an open interval outside the support of* $\hat{F}^{c_n,H_n}$ *for all large* n, *where* $H_n = F^{T_n}$.

Then

$$\mathsf{P}(\textit{no eigenvalue of } B_n \textit{ appears in } [a,b] \textit{ for all large } n) = 1.$$

Steps in proof: (1) Let $\underline{B}_n = (1/N)X_n^*T_nX_n$ $\underline{m}_n = m_{F^{\underline{B}_n}}$ and $\underline{m}_n^0 = m_{\hat{F}^{c_n,H_n}}$. Then for $z = x + iv_n$

$$\sup_{x\in[a,b]} |\underline{m}_n(z) - \underline{m}_n^0(z)| = o(1/Nv_n) \quad \text{a.s.}$$

when $v_n = N^{-1/68}$.

(2) The proof of (1) allows (1) to hold for $Im(z) = \sqrt{2}v_n, \sqrt{3}v_n, \ldots, \sqrt{34}v_n$. Then almost surely

$$\max_{k\in\{1,\ldots,34\}} \sup_{x\in[a,b]} |\underline{m}_n(x + i\sqrt{k}v_n) - \underline{m}_n^0(x + i\sqrt{k}v_n)| = o(v_n^{67}).$$

We take the imaginary part of these Stieltjes transforms and get

$$\max_{k\in\{1,2\ldots,34\}} \sup_{x\in[a,b]} \left| \int \frac{d(F^{\underline{B}_n}(\lambda) - \hat{F}^{c_n,H_n}(\lambda))}{(x-\lambda)^2 + kv_n^2} \right| = o(v_n^{66}) \quad \text{a.s.}$$

Upon taking differences we find with probability one

$$\max_{k_1\neq k_2} \sup_{x\in[a,b]} \left| \int \frac{v_n^2\, d(F^{\underline{B}_n}(\lambda) - \hat{F}^{c_n,H_n}(\lambda))}{((x-\lambda)^2 + k_1v_n^2)((x-\lambda)^2 + k_2v_n^2)} \right| = o(v_n^{66})$$

$$\max_{\substack{k_1,k_2,k_3\\ \text{distinct}}} \sup_{x\in[a,b]} \left| \int \frac{(v_n^2)^2\, d(F^{\underline{B}_n}(\lambda) - \hat{F}^{c_n,H_n}(\lambda))}{((x-\lambda)^2 + k_1v_n^2)((x-\lambda)^2 + k_2v_n^2)((x-\lambda)^2 + k_3v_n^2)} \right| = o(v_n^{66})$$

$$\vdots$$

$$\sup_{x\in[a,b]} \left| \int \frac{(v_n^2)^{33}\, d(F^{\underline{B}_n}(\lambda) - \hat{F}^{c_n,H_n}(\lambda))}{((x-\lambda)^2 + v_n^2)((x-\lambda)^2 + 2v_n^2)\cdots((x-\lambda)^2 + 34v_n^2)} \right| = o(v_n^{66}).$$

Thus with probability one

$$\sup_{x\in[a,b]}\left|\int \frac{d(F^{\underline{B}_n}(\lambda) - \hat{F}^{c_n,H_n}(\lambda))}{((x-\lambda)^2+v_n^2)((x-\lambda)^2+2v_n^2)\cdots((x-\lambda)^2+34v_n^2)}\right| = o(1)\,.$$

Let $0 < a' < a$, $b' > b$ be such that $[a',b']$ is also in the open interval outside the support of $\hat{F}^{c_n,H_n}$ for all large n. We split up the integral and get with probability one

$$\sup_{x\in[a,b]}\left|\int \frac{I_{[a',b']^c}(\lambda)\,d(F^{\underline{B}_n}(\lambda) - \hat{F}^{c_n,H_n}(\lambda))}{((x-\lambda)^2+v_n^2)((x-\lambda)^2+2v_n^2)\cdots((x-\lambda)^2+34v_n^2)}\right.$$

$$\left. + \sum_{\lambda_j\in[a',b']} \frac{v_n^{68}}{((x-\lambda_j)^2+v_n^2)((x-\lambda_j)^2+2v_n^2)\cdots((x-\lambda_j)^2+34v_n^2)}\right| = o(1).$$

Now if, for each term in a subsequence satisfying the above, there is at least one eigenvalue contained in $[a,b]$, then the sum, with x evaluated at these eigenvalues, will be uniformly bounded away from 0. Thus, at these same x values, the integral must also stay uniformly bounded away from 0. But the integral MUST converge to zero *a.s.* since the integrand is bounded and with probability one, both $F^{\underline{B}_n}$ and $\hat{F}^{c_n,H_n}$ converge weakly to the same limit having no mass on $\{a',b'\}$. Contradiction!

The last result is on the rate of convergence of linear statistics of the eigenvalues of B_n, that is, quantities of the form

$$\int f(x)dF^{B_n}(x) = \frac{1}{n}\sum_{i=1}^{n} f(\lambda_i)$$

where f is a function defined on $[0,\infty)$, and the λ_i's are the eigenvalues of B_n. The result establishes the rate to be $1/n$ for analytic f. It considers integrals of functions with respect to

$$G_n(x) = n[F^{B_n}(x) - F^{c_n,H_n}(x)]$$

where for any $d > 0$ and d.f. G, $F^{d,G}$ is the limiting e.d.f. of $B_n = (1/N)T_n^{1/2}X_nX_n^*T_n^{1/2}$ corresponding to limiting ratio d and limiting F^{T_n} G.

Theorem 7.2. ([2]). *Under the assumptions in Theorem 7.1, Let $f_1,\ldots,f_r$ be C^1 functions on $\mathbb{R}$ with bounded derivatives, and analytic on an open interval containing*

$$[\liminf_n \lambda_{\min}^{T_n} I_{(0,1)}(c)(1-\sqrt{c})^2, \limsup_n \lambda_{\max}^{T_n}(1+\sqrt{c})^2].$$

Let $\underline{m} = m_{\underline{\hat{F}}}$. *Then*

(1) *the random vector*

$$\left(\int f_1(x)\, dG_n(x), \ldots, \int f_r(x)\, dG_n(x)\right) \tag{7.1}$$

forms a tight sequence in n.

(2) *If* X_{11} *and* T_n *are real and* $\mathsf{E}(X_{11}^4) = 3$, *then (7.1) converges weakly to a Gaussian vector* $(X_{f_1}, \ldots, X_{f_r})$, *with means*

$$\mathsf{E}X_f = -\frac{1}{2\pi i}\int f(z) \frac{c\int \frac{\underline{m}(z)^3 t^2 dH(t)}{(1+t\underline{m}(z))^3}}{\left(1 - c\int \frac{\underline{m}(z)^2 t^2 dH(t)}{(1+t\underline{m}(z))^2}\right)^2} dz \tag{7.2}$$

and covariance function

$$Cov(X_f, X_g) = -\frac{1}{2\pi^2}\iint \frac{f(z_1)g(z_2)}{(\underline{m}(z_1) - \underline{m}(z_2))^2} \frac{d}{dz_1}\underline{m}(z_1) \frac{d}{dz_2}\underline{m}(z_2) dz_1 dz_2 \tag{7.3}$$

$(f, g \in \{f_1, \ldots, f_r\})$. *The contours in (7.2) and (7.3) (two in (7.3) which we may assume to be non-overlapping) are closed and are taken in the positive direction in the complex plane, each enclosing the support of* $F^{c,H}$.

(3) *If* X_{11} *is complex with* $\mathsf{E}(X_{11}^2) = 0$ *and* $\mathsf{E}(|X_{11}|^4) = 2$, *then (2) also holds, except the means are zero and the covariance function is 1/2 the function given in (7.3).*

(4) *If the assumptions in (2) or (3) were to hold, then* G_n, *considered as a random element in* $D[0, \infty)$ *(the space of functions on* $[0, \infty)$ *right-continuous with left-hand limits, together with the Skorohod metric) cannot form a tight sequence in* $D[0, \infty)$.

The proof relies on the identity

$$\int f(x) dG(x) = -\frac{1}{2\pi i}\int f(z) m_G(z) dz$$

(f analytic on the support of G, contour positively oriented around the support), and establishes the following results on

$$M_n(z) = n[m_{F^{B_n}}(z) - m_{F^{c_n, H_n}}(z)].$$

(a) $\{M_n(z)\}$ forms a tight sequence for z in a sufficiently large contour about the origin.

(b) If X_{11} is complex with $\mathsf{E}(X_{11}^2) = 0$ and $\mathsf{E}(X_{11}^4) = 2$, then for $z_1, \ldots, z_r$ with nonzero imaginary parts,

$$(\text{Re } M_n(z_1), \text{ Im } M_n(z_1), \ldots, \text{Re } M_n(z_r), \text{ Im } M_n(z_r))$$

converges weakly to a mean zero Gaussian vector. It follows that M_n, viewed as a random element in the metric space of continuous $\mathbb{R}^2$-valued functions with domain restricted to a contour in the complex plane, converges weakly to a (2 dimensional) Gaussian process M. The limiting covariance function can be derived from the formula

$$\mathsf{E}(M(z_1)M(z_2)) = \frac{\underline{m}'(z_1)\underline{m}'(z_2)}{(\underline{m}(z_1) - \underline{m}(z_2))^2} - \frac{1}{(z_1 - z_2)^2}.$$

(c) If X_{11} is real and $\mathsf{E}(X_{11}^4) = 3$ then (b) still holds, except the limiting mean can be derived from

$$\mathsf{E}M(z) = \frac{c \int \frac{\underline{m}^3 t^2 dH(t)}{(1+t\underline{m})^3}}{\left(1 - c \int \frac{\underline{m}^2 t^2 dH(t)}{(1+t\underline{m})^2}\right)^2}$$

and "covariance function" is twice that of the above function.

The difference between (2) and (3), and the difficulty in extending beyond these two cases, arise from

$$\mathsf{E}(X_{\cdot 1}^* A X_{\cdot 1} - \operatorname{tr} A)(X_{\cdot 1}^* B X_{\cdot 1} - \operatorname{tr} B)$$

$$= (\mathsf{E}(|X_{11}|^4) - |\mathsf{E}(X_{11}^2)|^2 - 2) \sum_i a_{ii} b_{ii} + |\mathsf{E}(X_{11}^2)|^2 \operatorname{tr} AB^T + \operatorname{tr} AB,$$

valid for square matrices A and B.

One can show

$$(7.2) = \frac{1}{2\pi} \int f'(x) \arg\left(1 - c \int \frac{t^2 \underline{m}^2(x)}{(1 + t\underline{m}(x))^2} dH(t)\right) dx$$

and

$$(7.3) = \frac{1}{\pi^2} \iint f'(x) g'(y) \ln \left| \frac{\underline{m}(x) - \overline{\underline{m}}(y)}{\underline{m}(x) - \underline{m}(y)} \right| dx dy$$

$$= \frac{1}{2\pi^2} \iint f'(x) g'(y) \ln \left(1 + 4 \frac{\underline{m}_i(x) \underline{m}_i(y)}{|\underline{m}(x) - \underline{m}(y)|^2}\right) dx dy$$

where $\underline{m}_i = \Im \underline{m}$.

For case (2) with $H = I_{[1,\infty)}$ we have for $f(x) = \ln x$ and $c \in (0,1)$

$$\mathsf{E}X_{\ln} = \frac{1}{2} \ln(1 - c) \qquad \text{and} \qquad \operatorname{Var} X_{\ln} = -2 \ln(1 - c).$$

Also, for $c > 0$

$$\mathsf{E}X_{x^r} = \frac{1}{4}((1-\sqrt{c})^{2r} + (1+\sqrt{c})^{2r}) - \frac{1}{2}\sum_{j=0}^{r}\binom{r}{j}^2 c^j$$

and

$$\operatorname{Cov}(X_{x^{r_1}}, X_{x^{r_2}}) = 2c^{r_1+r_2}\sum_{k_1=0}^{r_1-1}\sum_{k_2=0}^{r_2}\binom{r_1}{k_1}\binom{r_2}{k_2}\left(\frac{1-c}{c}\right)^{k_1+k_2}$$

$$\times \sum_{\ell=1}^{r_1-k_1}\ell\binom{2r_1-1-(k_1+\ell)}{r_1-1}\binom{2r_2-1-k_2+\ell}{r_2-1}.$$

(see [5]).

References

1. Z. D. Bai and J. W. Silverstein, No eigenvalues outside the support of the limiting spectral distribution of large-dimensional sample covariance matrices, *Ann. Probab.* **26**(1) (1998) 316–345.
2. Z. D. Bai and J. W. Silverstein, CLT for linear spectral statistics of large dimensional sample covariance matrices, *Ann. Probab.* **32**(1A) (2004) 553–605.
3. R. B. Dozier and J. W. Silverstein, On the empirical distribution of eigenvalues of large dimensional information-plus-noise type matrices, *J. Multivariate Anal.* **98**(4) (2007) 678–694.
4. R. B. Dozier and J. W. Silverstein, Analysis of the limiting spectral distribution of large dimensional information-plus-noise type matrices, *J. Multivariate Anal.* **98**(6) (2007) 1099–1122.
5. D. Jonsson, Some limit theorems for the eigenvalues of a sample covariance matrix, *J. Multivariate Anal.* **12**(1) (1982) 1–38.
6. V. A. Marčenko and L. A. Pastur, Distribution of eigenvalues for some sets of random matrices, *Math. USSR-Sb.* **1** (1967) 457–483.
7. J. W. Silverstein, Strong convergence of the empirical distribution of eigenvalues of large dimensional random matrices, *J. Multivariate Anal.* **55**(2) (1995) 331–339.
8. J. W. Silverstein and Z. D. Bai, On the empirical distribution of eigenvalues of a class of large dimensional random matrices, *J. Multivariate Anal.* **54**(2) (1995) 175–192.
9. J. W. Silverstein and S. I. Choi, Analysis of the limiting spectral distribution function of large dimensional random matrices, *J. Multivariate Anal.* **54**(2) (1995) 295–309.
10. Y. Q. Yin, Limiting spectral distribution for of random matrices, *J. Multivariate Anal.* **20** (1986) 50–68.

BETA RANDOM MATRIX ENSEMBLES

Peter J. Forrester
Department of Mathematics and Statistics
University of Melbourne, Victoria 3010, Australia
E-mail: p.forrester@ms.unimelb.edu.au

In applications of random matrix theory to physics, time reversal symmetry implies one of three exponents $\beta = 1, 2$ or 4 for the repulsion s^β between eigenvalues at spacing s, $s \to 0$. However, in the corresponding eigenvalue probability density functions (p.d.f.'s), β is a natural positive real variable. The general β p.d.f.'s have alternative physical interpretations in terms of certain classical log-potential Coulomb systems and quantum many body systems. Each of these topics is reviewed, along with the problem of computing correlation functions for general β. There are also random matrix constructions which give rise to these general β p.d.f.'s. An inductive approach to the latter topic using explicit formulas for the zeros of certain random rational functions is given.

1. Introduction

1.1. *Log-gas systems*

In equilibrium classical statistical mechanics, the control variables are the absolute temperature T and the particle density ρ. The state of a system can be calculated by postulating that the probability density function (p.d.f.) for the event that the particles are at positions $\vec{r}_1, \ldots, \vec{r}_N$ is proportional to the Boltzmann factor $e^{-\beta U(\vec{r}_1,\ldots,\vec{r}_N)}$. Here $U(\vec{r}_1, \ldots, \vec{r}_N)$ denotes the total potential energy of the system, while $\beta := 1/k_B T$ with k_B denoting Boltzmann's constant is essentially the inverse temperature. Then for the system confined to a domain Ω, the canonical average of any function $f(\vec{r}_1, \ldots, \vec{r}_N)$ (for example the energy itself) is given by

$$\langle f \rangle := \frac{1}{\hat{Z}_N} \int_\Omega d\vec{r}_1 \cdots \int_\Omega d\vec{r}_N \, f(\vec{r}_1, \ldots, \vec{r}_N) e^{-\beta U(\vec{r}_1,\ldots,\vec{r}_N)},$$

where

$$\hat{Z}_N = \int_I d\vec{r}_1 \cdots \int_I d\vec{r}_N e^{-\beta U(r_1,\dots,r_N)}. \tag{1.1}$$

In the so called thermodynamic limit, $N, |\Omega| \to \infty$, $N/|\Omega| = \rho$ fixed, such averages can display non-analytic behavior indicative of a phase transition.

The most common situation is when the potential energy U consists of a sum of one body and two body terms,

$$U(\vec{r}_1, \dots, \vec{r}_N) = \sum_{j=1}^{N} V_1(\vec{r}_j) + \sum_{1 \le j < k \le N} V_2(|\vec{r}_k - \vec{r}_j|).$$

Our interest is when the pair potential V_2 is logarithmic,

$$V_2(|\vec{r}_k - \vec{r}_j|) = -\log|\vec{r}_k - \vec{r}_j|.$$

Physically this is the law of repulsion between two infinitely long wires, which can be thought of as two-dimensional charges. Because the charges are of the same sign, with strength taken to be unity, without the presence of a confining potential they would repel to infinity. Indeed, well defined thermodynamics requires that the system be overall charge neutral. This can be achieved by immersing the charges in a smeared out neutralizing density. In particular, let this density have profile $\rho_b(\vec{r})$. It is thus required that

$$\int_\Omega \rho_b(\vec{r})\, d\vec{r} = -N$$

while the one body potential V_1 is calculated according to

$$V_1(\vec{r}) = \int_\Omega \log|\vec{r} - \vec{r}_j| \rho_b(\vec{r})\, d\vec{r}. \tag{1.2}$$

Consider first some specific log-potential systems (log-gases for short) confined to one-dimensional domains. Four distinct examples are relevant. These are the system on the real line, with the neutralizing background having profile

$$\rho_b(x) = \frac{\sqrt{2N}}{\pi}\sqrt{1 - \frac{x^2}{2N}}, \qquad |x| < \sqrt{2N}; \tag{1.3}$$

the system on a half line with image charges of the same sign in $x < 0$, a fixed charge of strength $(a-1)/2$ at the origin, and a neutralizing background charge density in $x > 0$ with profile

$$\rho_b(x) = \frac{\sqrt{4N}}{\pi}\sqrt{1 - \frac{x^2}{4N}}, \qquad 0 < x < 2\sqrt{N}; \tag{1.4}$$

the system on the unit interval $[-1,1]$ with fixed charges of strengths $(a-1)/2+1/\beta$ and $(b-1)/2+1/\beta$ at $y=1$ and $y=-1$ respectively and a neutralizing background

$$\rho_b(x) = \frac{N}{\pi\sqrt{1-x^2}}, \qquad |x|<1; \tag{1.5}$$

and the system on a unit circle with a uniform neutralizing background density.

Physically one expects charged systems to be locally charge neutral. Accepting this, the particle densities must then, to leading order, coincide with the background densities. In the above examples, this implies that in the bulk of the system the particle densities are dependent on N, which in turn means that there is not yet a well defined thermodynamic limit. To overcome this, note the special property of the logarithmic potential that it is unchanged, up to an additive constant, by the scaling of the coordinates $x_j \mapsto cx_j$. Effectively the density is therefore not a control variable, as it determines only the length scale in the logarithmic potential.

Making use of (1.2), for the four systems the total energy of the system can readily be computed (see [13] for details) to give the corresponding Boltzmann factors. They are proportional to

$$\prod_{j=1}^N e^{-(\beta/2)x_j^2} \prod_{1\le j<k\le N} |x_k-x_j|^\beta \tag{1.6}$$

$$\prod_{j=1}^N |x_j|^{\beta\alpha} e^{-\beta x_j^2/2} \prod_{1\le j<k\le N} |x_k^2-x_j^2|^\beta \tag{1.7}$$

$$\prod_{j=1}^N (1-x_j)^{a\beta/2}(1+x_j)^{b\beta/2} \prod_{1\le j<k\le N} |x_k-x_j|^\beta \tag{1.8}$$

$$\prod_{1\le j<k\le N} |e^{i\theta_k}-e^{i\theta_j}|^\beta \tag{1.9}$$

respectively. We remark that changing variables $x_j=\cos\theta_j$, $0<\theta_j<\pi$ in (1.8) the Boltzmann factor becomes proportional to

$$\prod_{j=1}^N (\sin\theta_j/2)^{(a-1/2)\beta+1/2}(\cos\theta_j/2)^{(b-1/2)\beta+1/2}(\sin\theta_j)^{\beta/2}$$
$$\times \prod_{1\le j<k\le N} |\sin((\theta_k-\theta_j)/2)\sin((\theta_k+\theta_j)/2)|^\beta . \tag{1.10}$$

This corresponds to the log-gas on a half circle $0 \le \theta_j \le \pi$, with image charges of the same sign at $2\pi - \theta_j$, a charge of strength $(a - 1/2) + 1/\beta$ at $\theta = 0$, and a charge of strength $(b - 1/2) + 1/\beta$ at $\theta = \pi$.

1.2. *Quantum many body systems*

The setting of Section 1.1 was equilibrium statistical mechanics. As a generalization, let us suppose now that there are dynamics due to Brownian motion in a (fictitious) viscous fluid with friction coefficient γ. The evolution of the p.d.f. $p_\tau(x_1, \dots, x_N)$ for the joint density of the particles is given by the solution of the Fokker-Planck equation (see e.g. [32])

$$\gamma \frac{\partial p_\tau}{\partial \tau} = \mathcal{L} p_\tau \qquad \text{where} \quad \mathcal{L} = \sum_{j=1}^N \frac{\partial}{\partial \lambda_j} \left(\frac{\partial U}{\partial \lambda_j} + \beta^{-1} \frac{\partial}{\partial \lambda_j} \right). \tag{1.11}$$

In general the steady state solution of this equation is the Boltzmann factor $e^{-\beta U}$,

$$\mathcal{L} e^{-\beta U} = 0. \tag{1.12}$$

Another general property is the operator identity

$$e^{\beta U/2} \mathcal{L} e^{-\beta U/2} = \sum_{j=1}^N \left(\frac{1}{\beta} \frac{\partial^2}{\partial x_j^2} - \frac{\beta}{4} \left(\frac{\partial U}{\partial x_j} \right)^2 + \frac{1}{2} \frac{\partial^2 U}{\partial x_j^2} \right), \tag{1.13}$$

relating $\mathcal{L}$ to an Hermitian operator.

For the potentials implied by the Boltzmann factors (1.6), (1.7), (1.9) and (1.10) the conjugation (1.13) gives

$$-e^{\beta U/2} \mathcal{L} e^{-\beta U/2} = (H - E_0)/\beta \tag{1.14}$$

where H is a Schrödinger operator consisting of one and two body terms only,

$$H = -\sum_{j=1}^N \frac{\partial^2}{\partial x_j^2} + \sum_{j=1}^N v_1(x_j) + \sum_{1 \le j < k \le N} v_2(x_j, x_k).$$

Explicitly, one finds [2]

$$H = H^{(H)} := -\sum_{j=1}^{N} \frac{\partial^2}{\partial x_j^2} + \frac{\beta^2}{4}\sum_{j=1}^{N} x_j^2 + \beta(\beta/2-1) \sum_{1\le j<k\le N} \frac{1}{(x_j-x_k)^2} \tag{1.15}$$

$$H = H^{(L)} := -\sum_{j=1}^{N} \frac{\partial^2}{\partial x_j^2} + \sum_{j=1}^{N}\Big(\frac{\beta a'}{2}\Big(\frac{\beta a'}{2}-1\Big)\frac{1}{x_j^2} - \frac{\beta^2}{4}x_j^2\Big) + \beta(\beta/2-1) \sum_{1\le j<k\le N} \Big(\frac{1}{(x_k-x_j)^2} + \frac{1}{(x_k+x_j)^2}\Big) \tag{1.16}$$

$$H = H^{(C)} := -\sum_{j=1}^{N} \frac{\partial^2}{\partial \theta_j^2} + (\beta/4)(\beta/2-1) \sum_{1\le j<k\le N} \frac{1}{\sin^2((\theta_k-\theta_j)/2)} \tag{1.17}$$

$$H = H^{(J)} := -\sum_{j=1}^{N} \frac{\partial^2}{\partial \phi_j^2} + \sum_{j=1}^{N}\Big(\frac{a'\beta}{2}\Big(\frac{a'\beta}{2}-1\Big)\frac{1}{\sin^2\phi_j} + \frac{b'\beta}{2}\Big(\frac{b'\beta}{2}-1\Big)\frac{1}{\cos^2\phi_j}\Big) + \beta(\beta/2-1) \sum_{1\le j<k\le N} \Big(\frac{1}{\sin^2(\phi_j-\phi_k)} + \frac{1}{\sin^2(\phi_j+\phi_k)}\Big) \tag{1.18}$$

where $a' = a+1/\beta$, $b' = b+1/\beta$. Thus all the pair potentials are proportional to $1/r^2$, where r is the separation between particles or between particles and their images. Such quantum many body systems were first studied by Calogero [5] and Sutherland [35].

It follows from (1.12) and (1.14) that $e^{-\beta U/2}$ is an eigenfunction of H with eigenvalue E_0. Since $e^{-\beta U/2}$ is non-negative, it must in fact be the ground state. This suggests considering a conjugation of the Schrödinger operators with respect to this eigenfunction. Consider for definiteness (1.17). A direct computation gives

$$\tilde{H}^{(C)} := e^{\beta U/2}(H^{(C)} - E_0)e^{-\beta U/2} = \sum_{j=1}^{N}\Big(z_j\frac{\partial}{\partial z_j}\Big)^2 + \frac{N-1}{\alpha} + \frac{2}{\alpha}\sum_{1\le j<k\le N} \frac{z_j z_k}{z_j-z_k}\Big(\frac{\partial}{\partial z_j} - \frac{\partial}{\partial z_k}\Big) \tag{1.19}$$

where $z_j := e^{ix_j}$ and $\alpha = 2/\beta$. This operator admits a complete set of symmetric polynomial eigenfunctions $P_\kappa(z) = P_\kappa(z_1,\dots,z_N;\alpha)$ labeled by partitions $\kappa := (\kappa_1,\dots,\kappa_N)$, $\kappa_1 \ge \cdots \ge \kappa_N \ge 0$, known as the symmetric Jack polynomials [27, 13]. These polynomials have the structure

$$P_\kappa(z) = m_\kappa + \sum_{\sigma<\kappa} b_{\kappa\sigma} m_\sigma$$

where m_κ denotes the monomial symmetric function in the variables $z_1,\dots,z_N$ associated with the partition κ (for example, with $\kappa = 21$ and $N = 2$, $m_{21} = z_1^2 z_2 + z_1 z_2^2$), $<$ denotes the dominance ordering for partitions, and the coefficients $b_{\kappa\sigma}$ are independent of N.

1.3. *Selberg correlation integrals*

The symmetric Jack polynomials can be used as a basis to define a class of generalized hypergeometric functions, which in turn have direct relevance to the calculation of correlation functions in the log-gas. To set up the definition of the former, introduce the generalized factorial function

$$[u]_\kappa^{(\alpha)} := \prod_{j=1}^N \frac{\Gamma(u - \frac{1}{\alpha}(j-1) + \kappa_j)}{\Gamma(u - \frac{1}{\alpha}(j-1))}, \tag{1.20}$$

the quantity

$$d'_\kappa := \prod_{(i,j)\in\kappa} \Big(\alpha(\kappa_i - j + 1) + (\kappa'_j - i)\Big) \tag{1.21}$$

(here κ'_j denotes the length of column j in the diagram of κ), and the renormalized Jack polynomial

$$C_\kappa^{(\alpha)}(x) := \frac{\alpha^{|\kappa|}|\kappa|!}{d'_\kappa} P_\kappa(x;\alpha). \tag{1.22}$$

In terms of these quantities, the generalized hypergeometric functions ${}_pF_q^{(\alpha)}$ are specified by

$$\begin{aligned} &{}_pF_q^{(\alpha)}(a_1,\dots,a_p,b_1,\dots,b_q;x_1,\dots,x_m) \\ &\quad := \sum_\kappa \frac{1}{|\kappa|!} \frac{[a_1]_\kappa^{(\alpha)} \cdots [a_p]_\kappa^{(\alpha)}}{[b_1]_\kappa^{(\alpha)} \cdots [b_q]_\kappa^{(\alpha)}} C_\kappa^{(\alpha)}(x_1,\dots,x_m). \end{aligned} \tag{1.23}$$

Since in the one-variable case we have $\kappa = k$, $C_k^{(\alpha)}(x) = x^k$ and $[u]_k^{(\alpha)} = (u)_k$, we see that with $m = 1$ the generalized hypergeometric function ${}_pF_q^{(\alpha)}$ reduces to the classical hypergeometric function ${}_pF_q$.

There are two cases in which ${}_pF_q^{(\alpha)}$ can be expressed in terms of elementary functions [24, 37]. These are the generalized binomial theorem

$$ {}_1F_0^{(\alpha)}(a; x_1, \dots, x_m) = \prod_{j=1}^{m}(1 - x_j)^{-a} \tag{1.24} $$

and its limiting form

$$ {}_0F_0^{(\alpha)}(x_1, \dots, x_m) = e^{x_1 + \cdots + x_m}. \tag{1.25} $$

The latter can be deduced from the confluence relation

$$ \begin{aligned} &\lim_{a_p \to \infty} {}_pF_q^{(\alpha)}(a_1, \dots, a_p; b_1, \dots, b_q; x_1/a_p, \dots, x_m/a_p) \\ &\quad = {}_{p-1}F_q^{(\alpha)}(a_1, \dots, a_{p-1}; b_1, \dots, b_q; x_1, \dots, x_m), \end{aligned} \tag{1.26} $$

which follows from the explicit form (1.20) of $[a_p]_\kappa^{(\alpha)}$ and the fact that $C_\kappa^{(\alpha)}(x)$ is homogeneous of degree $|\kappa|$.

We will now relate ${}_2F_1^{(\alpha)}$ to a generalization of the Selberg integral referred to as a Selberg correlation integral. First we recall that the Selberg integral is the multidimensional generalization of the beta integral

$$ S_N(\lambda_1, \lambda_2, \lambda) := \int_0^1 dx_1 \cdots \int_0^1 dx_N \prod_{l=1}^{N} x_l^{\lambda_1}(1 - x_l)^{\lambda_2} \prod_{1 \le j < k \le N} |x_k - x_j|^{2\lambda}. $$

This can be transformed to the trigonometric form

$$ \begin{aligned} M_N(a, b, \lambda) := &\int_{-1/2}^{1/2} d\theta_1 \cdots \int_{-1/2}^{1/2} d\theta_N \prod_{l=1}^{N} e^{\pi i \theta_l (a-b)} |1 + e^{2\pi i \theta_l}|^{a+b} \\ &\times \prod_{1 \le j < k \le N} |e^{2\pi i \theta_k} - e^{2\pi i \theta_j}|^{2\lambda} \end{aligned} $$

known as the Morris integral. The Selberg correlation integral refers to the generalizations

$$ \begin{aligned} S_N(t_1, \dots, t_m; \lambda_1, \lambda_2, 1/\alpha) := &\frac{1}{S_N(\lambda_1, \lambda_2, 1/\alpha)} \int_0^1 dx_1 \cdots \int_0^1 dx_N \\ &\times \prod_{l=1}^{N} x_l^{\lambda_1}(1 - x_l)^{\lambda_2} \prod_{l'=1}^{m}(1 - t_{l'} x_l) \prod_{j<k} |x_j - x_k|^{2/\alpha}, \end{aligned} $$

$$ \begin{aligned} \tilde{S}_N(t_1, \dots, t_m; \lambda_1, \lambda_2, 1/\alpha) := &\frac{1}{S_N(\lambda_1 + m, \lambda_2, 1/\alpha)} \int_0^1 dx_1 \cdots \int_0^1 dx_N \\ &\times \prod_{l=1}^{N} x_l^{\lambda_1}(1 - x_l)^{\lambda_2} \prod_{l'=1}^{m}(t_{l'} - x_l) \prod_{j<k} |x_j - x_k|^{2/\alpha}, \end{aligned} $$

and its trigonometric form

$$M_N(t_1,\dots,t_m;a,b,1/\alpha) := \frac{1}{M_N(a,b,1/\alpha)}\int_{-1/2}^{1/2} dx_1 \cdots \int_{-1/2}^{1/2} dx_N$$
$$\times \prod_{l=1}^{N} e^{\pi i x_l (a-b)} |1+e^{2\pi i x_l}|^{a+b} \prod_{l'=1}^{m} (1+t_{l'} e^{2\pi i x_l}) \prod_{j<k} |e^{2\pi i x_k} - e^{2\pi i x_j}|^{2/\alpha}.$$

These have the generalized hypergeometric function evaluations [24, 14]

$$S_N(t_1,\dots,t_m;\lambda_1,\lambda_2,1/\alpha) = {}_2F_1^{(1/\alpha)}(-N, -(N-1)-\alpha(\lambda_1+1); \\ -2(N-1)-\alpha(\lambda_1+\lambda_2+2); t_1,\dots,t_m), \tag{1.27}$$

$$\tilde{S}_N(t_1,\dots,t_m;\lambda_1,\lambda_2,1/\alpha) = {}_2F_1^{(1/\alpha)}(-N, \\ (N-1)+\alpha(\lambda_1+\lambda_2+m+1); \alpha(\lambda_1+m); t_1,\dots,t_m), \tag{1.28}$$

and

$$M_N(t_1,\dots,t_m;a,b,1/\alpha) = {}_2F_1^{(1/\alpha)}(-N, \alpha b; -(N-1)-\alpha(1+a); t_1,\dots,t_m). \tag{1.29}$$

On the other hand, the generalized binomial expansion allows the generalized hypergeometric function ${}_2F_1^{(\alpha)}$ in m variables to be expressed as an m-dimensional integral, provided all the arguments are equal. Thus we have [18, 15]

$$\frac{1}{M_N(a,b,1/\alpha)}\int_{-1/2}^{1/2} dx_1 \cdots \int_{-1/2}^{1/2} dx_N \prod_{l=1}^{N} e^{\pi i x_l(a-b)} |1+e^{2\pi i x_l}|^{a+b} \\ \times (1+te^{2\pi i x_l})^{-r} \prod_{1\le j<k\le N} |e^{2\pi i x_k} - e^{2\pi i x_j}|^{2/\alpha} \\ = {}_2F_1^{(\alpha)}(r,-b;\frac{1}{\alpha}(N-1)+a+1;t_1,\dots,t_N)\Big|_{t_1=\cdots=t_N=t}, \tag{1.30}$$

$$\frac{1}{S_N(\lambda_1,\lambda_2,1/\alpha)}\int_0^1 dx_1 \cdots \int_0^1 dx_N \prod_{l=1}^{N} x_l^{\lambda_1}(1-x_l)^{\lambda_2}(1-tx_l)^{-r} \\ \prod_{j<k} |x_j - x_k|^{2/\alpha} = {}_2F_1^{(\alpha)}(r, \frac{1}{\alpha}(N-1)+\lambda_1+1; \\ \frac{2}{\alpha}(N-1)+\lambda_1+\lambda_2+2; t_1,\dots,t_N)\Big|_{t_1=\cdots=t_N=t}. \tag{1.31}$$

In using (1.30) and (1.31) in (1.27) and (1.29) it may happen that the parameters are such that the former are divergent. To overcome this, use can be made of certain transformation formulas satisfied by the ${}_2F_1^{(\alpha)}$. One such formula, which is restricted to cases in which the series (1.23) terminates, is [14]

$$
\begin{aligned}
&{}_2F_1^{(\alpha)}(a,b;c;t_1,\dots,t_m) \\
&\quad = \frac{{}_2F_1^{(\alpha)}(a,b;a+b+1+(m-1)/\alpha-c;1-t_1,\dots,1-t_m)}{{}_2F_1^{(\alpha)}(a,b;a+b+1+(m-1)/\alpha-c;t_1,\dots,t_m)\Big|_{t_1=\cdots=t_m=1}}. \qquad (1.32)
\end{aligned}
$$

Another, which generalizes one of the classical Kummer relations, reads [37]

$$
\begin{aligned}
&{}_2F_1^{(\alpha)}(a,b;c;t_1,\dots,t_m) \\
&\quad = \prod_{j=1}^{m}(1-t_j)^{-a}\,{}_2F_1^{(\alpha)}\Big(a,c-b;c;-\frac{t_1}{1-t_1},\dots,-\frac{t_m}{1-t_m}\Big). \qquad (1.33)
\end{aligned}
$$

1.4. *Correlation functions*

For a one-dimensional statistical mechanical system with Boltzmann factor $e^{-\beta U}$ confined to a domain I, the n-particle correlation function is defined by

$$
\rho_{(n)}(\vec{r}_1,\dots,\vec{r}_n) = \frac{N(N-1)\cdots(N-n+1)}{\hat{Z}_N}\int_I dr_{n+1}\cdots\int_I dr_N\, e^{-\beta U(r_1,\dots,r_N)} \qquad (1.34)
$$

where $\hat{Z}_N$ is specified by (1.1). In the case of the log-gas systems specified in Section 1.1 with even β, the Selberg correlation integral evaluations (1.28), (1.29) allows this to be expressed in terms of generalized hypergeometric functions.

Consider first the Boltzmann factor (1.8), but with the change of variables $x_j \mapsto 1-2x_j$ so that now $0<x_j<1$. Further set $a\beta/2=\lambda_1$ and $b\beta/2=\lambda_2$. The resulting p.d.f. is said to define the Jacobi β-ensemble. According to (1.34), with $N+n$ particles the corresponding n-point correlation is given by

$$
\begin{aligned}
\rho_{(n)}(r_1,\dots,r_n) &:= \frac{(N+n)_n}{S_{N+n}(\lambda_1,\lambda_2,\beta/2)}\prod_{k=1}^{n} r_k^{\lambda_1}(1-r_k)^{\lambda_2}\prod_{j<k}^{n}|r_k-r_j|^{\beta} \\
&\times\int_{[0,1]^N} dx_1\dots dx_N \prod_{j=1}^{N}\Big(x_j^{\lambda_1}(1-x_j)^{\lambda_2}\prod_{k=1}^{n}|x_j-r_k|^{\beta}\Big)\prod_{1\le j<k\le N}|x_k-x_j|^{\beta}. \qquad (1.35)
\end{aligned}
$$

In the case β even, the factor $|x_j - r_k|^\beta$ is a polynomial. Making use of (1.28) it follows that

$$\rho_{(n)}(r_1, \dots, r_n)$$
$$= (N+n)_n \frac{S_N(\lambda_1 + n\beta, \lambda_2, \beta/2)}{S_{N+n}(\lambda_1, \lambda_2, \beta/2)} \prod_{k=1}^n t_k^{\lambda_1}(1-t_k)^{\lambda_2} \prod_{j<k}^n |t_k - t_j|^\beta$$
$$\times {}_2F_1^{(\beta/2)}(-N, 2(\lambda_1+\lambda_2+m+1)/\beta + N - 1; 2(\lambda_1+m)/\beta; t_1, \dots, t_{\beta n}) \tag{1.36}$$

where

$$t_k = r_j \quad \text{for} \quad k = 1 + (j-1)\beta, \dots, j\beta \quad (j = 1, \dots, n). \tag{1.37}$$

For $n = 1$ the arguments (1.37) are equal, and we have available the β-dimensional integral representation (1.30). The get a convergent integral we must first apply the Kummer type transformation (1.33). Doing this gives [13]

$$\rho_{(1)}(r) = (N+1) \frac{S_N(\lambda_1 + \beta, \lambda_2, \beta/2)}{S_{N+1}(\lambda_1, \lambda_2, \beta/2)}$$
$$\times \frac{r^{\lambda_1}(1-r)^{\lambda_2}}{M_\beta(2(\lambda_1+1)/\beta - 1, 2(\lambda_2+1)/\beta + N - 1; 2/\beta)}$$
$$\times \int_{-1/2}^{1/2} dx_1 \cdots \int_{-1/2}^{1/2} dx_\beta \prod_{l=1}^\beta e^{\pi i x_l (2(\lambda_1 - \lambda_2)/\beta)} |1 + e^{2\pi i x_l}|^{2(\lambda_1+\lambda_2+2)/\beta + N - 1}$$
$$\times (e^{-\pi i x_l} - \frac{r}{1-r} e^{\pi i x_l})^N \prod_{1 \le j < k \le \beta} |e^{2\pi i x_k} - e^{2\pi i x_j}|^{4/\beta}. \tag{1.38}$$

The Boltzmann factor (1.7) with the change of variable $x_j^2 \mapsto x_j$, and $\alpha\beta = a\beta + 1$ is said to specify the Laguerre β-ensemble. It can be obtained from the Jacobi β-ensemble by the change of variables and limiting procedure

$$x_j \mapsto x_j / L, \quad \lambda_2 \mapsto L\beta/2, \quad \lambda_1 \mapsto a\beta/2, \quad L \to \infty. \tag{1.39}$$

The confluence (1.26) allows the same limiting procedure to be applied to (1.36). Thus we obtain

$$\rho_{(n)}(r_1,\dots,r_n) = \frac{(N+n)_n}{W_{a,\beta,N+n}} \left(\prod_{j=1}^n r_j^{a\beta/2} e^{-\beta r_j/2} \right) \prod_{1\le j<k\le n} |r_k - r_j|^\beta$$
$$\times \int_{(0,\infty)^N} dx_1 \cdots dx_N \prod_{j=1}^N \left(x_j^{a\beta/2} e^{-\beta x_j/2} \prod_{k=1}^n |r_k - x_j|^\beta \right)$$
$$\times \prod_{j<k} |x_k - x_j|^\beta, \tag{1.40}$$

where

$$W_{a,\beta,N} = \int_0^\infty dx_1 \cdots \int_0^\infty dx_N \prod_{j=1}^N x_j^{\beta a/2} e^{-\beta x_j/2} \prod_{j<k} |x_k - x_j|^\beta.$$

Applying the limiting procedure to (1.38) gives for the one-point correlation (i.e. the particle density) with β even the β-dimensional integral representation

$$\rho_{(1)}(r) = (N+1) \frac{W_{a,\beta,N}}{W_{a+2,\beta,N+1}} \frac{r^{a\beta/2} e^{-\beta r/2}}{M_\beta(2/\beta - 1, N, \beta/2)} \int_{-1/2}^{1/2} dx_1 \cdots \int_{-1/2}^{1/2} dx_\beta$$
$$\times \prod_{l=1}^\beta e^{\pi i x_l (2/\beta - 1 - N)} |1 + e^{2\pi i x_l}|^{N+2/\beta-1} e^{-re^{2\pi i x_l}}$$
$$\times \prod_{j<k}^\beta |e^{2\pi i x_k} - e^{2\pi i x_j}|^{4/\beta}. \tag{1.41}$$

For the Laguerre β-ensemble, in addition to the correlations at even β being expressible in terms of the confluent hypergeometric function, one can give similar evaluations of the probability $E_\beta^{(N)}((0,s))$ that the interval $(0,s)$ is particle free,

$$E_\beta^{(N)}(0;(0,s))$$
$$= \frac{1}{W_{a,\beta,N}} \int_s^\infty dx_1 \cdots \int_s^\infty dx_N \prod_{j=1}^N e^{-\beta x_j/2} x_j^{\beta a/2} \prod_{j<k} |x_k - x_j|^\beta$$
$$= \frac{e^{-N\beta s/2}}{W_{a,\beta,N}} \int_0^\infty dx_1 \cdots \int_0^\infty dx_N \prod_{j=1}^N e^{-\beta x_j/2} (x_j + s)^{\beta a/2} \prod_{j<k} |x_k - x_j|^\beta \tag{1.42}$$

where the second equality follows by the change of variables $x_j \mapsto x_j + s$. Thus it follows by applying the limiting procedure (1.39) to (1.36) that for $\beta a/2 =: m \in \mathbb{Z}_{\geq 0}$ [17]

$$E_\beta^{(N)}(0;(0,s)) = e^{-\beta N s/2} \, {}_1F_1^{(\beta/2)}(-N; 2m/\beta; x_1, \dots, x_m)\Big|_{x_j=-s}. \tag{1.43}$$

A closely related quantity is the distribution of the particle closest to the origin,

$$\begin{aligned} p_\beta^{(N)}(0;s) &= -\frac{d}{ds} E_\beta^{(N)}(0;(0,s)) \\ &= \frac{N e^{-N\beta s/2}}{W_{a,\beta,N}} s^{\beta a/2} \int_0^\infty dx_1 \cdots \int_0^\infty dx_{N-1} \\ &\quad \times \prod_{j=1}^{N-1} x_j^\beta e^{-\beta x_j/2} (x_j + s)^{\beta a/2} \prod_{j<k} |x_k - x_j|^\beta \end{aligned} \tag{1.44}$$

where the second equality follows by differentiating the first equality in (1.42) and then changing variables $x_j \mapsto x_j + s$. This multidimensional integral can be evaluated in an analogous way to that in (1.42) to give [17]

$$\begin{aligned} p_\beta^{(N)}(0;s) &= N s^m e^{-\beta N s/2} \frac{W_{a+2,\beta,N-1}}{W_{a,\beta,N}} \\ &\quad \times {}_1F_1^{(\beta/2)}(-N+1; 2m/\beta + 2; x_1, \dots, x_m)\Big|_{x_j=-s}. \end{aligned} \tag{1.45}$$

For the Gaussian β-ensemble with $N+n$ particles

$$\begin{aligned} \rho_{(n)}(r_1, \dots, r_n) &= \frac{(N+n)_n}{G_{\beta,N}} \prod_{j=1}^n e^{-\beta r_j^2/2} \prod_{1\leq j<k\leq N} |r_k - r_j|^\beta \\ \times \int_{(-\infty,\infty)^N} dx_1 \cdots dx_N & \prod_{j=1}^N \left(e^{-\beta x_j^2/2} \prod_{k=1}^n |r_k - x_j|^\beta \right) \prod_{j<k} |x_k - x_j|^\beta \end{aligned}$$

where

$$G_{\beta,N} = \int_{(-\infty,\infty)^N} dx_1 \cdots dx_N \prod_{j=1}^N e^{-\beta x_j^2/2} \prod_{j<k} |x_k - x_j|^\beta.$$

The Gaussian β-ensemble can be obtained from the Jacobi β-ensemble on $[0,1]$ through the change of variables

$$x_j \mapsto \frac{1}{2}\left(1 - \sqrt{\frac{\beta}{2}} \frac{x_j}{L}\right), \qquad \lambda_1 = \lambda_2 = L^2$$

and taking the limit $L \to \infty$. Applying this to (1.38) gives for the one-point density [2]

$$\rho_{(1)}(r) = (N+1)\frac{G_{\beta,N}}{G_{\beta,N+1}}\frac{e^{-\beta r^2/2}}{\tilde{G}_\beta}\int_{(-\infty,\infty)^\beta} du_1\cdots du_\beta$$

$$\times \prod_{j=1}^{\beta}(iu_j+r)^N e^{-u_j^2}\prod_{1\le j<k\le\beta}|u_k-u_j|^{4/\beta} \tag{1.46}$$

where

$$\tilde{G}_\beta = \int_{(-\infty,\infty)^\beta} dx_1\cdots dx_\beta \prod_{j=1}^{\beta} e^{-x_j^2}\prod_{j<k}|x_k-x_j|^{4/\beta}.$$

The last case to consider is the circular β-ensemble. With $N+n$ particles the n-point correlation function is given by

$$\rho_{(n)}(r_1,\dots,r_n) = \frac{(N+n)_n}{L^n}\frac{((\beta/2)!)^{N+n}}{(\beta(N+n)/2)!}$$

$$\times \prod_{1\le j<k\le n}|e^{2\pi i r_k/L}-e^{2\pi i r_j/L}|^\beta I_{N,n}(\beta;r_1,\dots,r_n) \tag{1.47}$$

where

$$I_{N,n}(\beta;r_1,\dots,r_n) := \int_{[0,1]^N} dx_1\cdots dx_N \prod_{j=1}^{N}\prod_{k=1}^{n}|1-e^{2\pi i(x_j-r_k/L)}|^\beta$$

$$\times \prod_{1\le j<k\le N}|e^{2\pi i x_k}-e^{2\pi i x_j}|^\beta, \tag{1.48}$$

and the angles have been scaled so that the circumference length of the circle is equal to L. Use of (1.29) and the transformation formula (1.32) shows [14] that for β even

$$\rho_{(n)}(r_1,\dots,r_n) = \frac{(N+n)_n}{L^n}\frac{((\beta/2)!)^{N+n}}{(\beta(N+n)/2)!}\prod_{1\le j<k\le n}|e^{2\pi i r_k/L}-e^{2\pi i r_j/L}|^\beta$$

$$\times M_N(n\beta/2,n\beta/2,\beta/2)\prod_{k=2}^{n}e^{\pi i N\beta(r_k-r_1)/L}$$

$$\times\, {}_2F_1^{(\beta/2)}(-N,n;2n;1-t_1,\dots,1-t_{(n-1)\beta}) \tag{1.49}$$

where

$$t_k := e^{-2\pi i(r_j-r_1)/L}, \qquad k=1+(j-2)\beta,\dots,(j-1)\beta \quad (j=2,\dots,n). \tag{1.50}$$

For the circular ensemble the one-point function is a constant, due to translation invariance. The two-point function is therefore a function of the single variable $r_2 - r_1$ as is clear from (1.50). Moreover, use of (1.31) allows the generalized hypergeometric function in (1.49) to be written as a β-dimensional integral [15]

$$\begin{aligned}
&\rho_{(2)}(r_1, r_2) \\
&= \frac{(N+2)(N+1)}{L^2} \frac{(\beta N/2)!((\beta/2)!)^{N+2}}{(\beta(N+2)/2)!} \frac{M_N(n\beta/2, n\beta/2, \beta/2)}{S_\beta(1-2/\beta, 1-2/\beta, 2/\beta)} \\
&\quad \times (2\sin\pi(r_1 - r_2)/L)^\beta e^{-\pi i \beta N(r_1 - r_2)/L} \int_{[0,1]^\beta} du_1 \cdots du_\beta \prod_{j=1}^{\beta} \\
&\quad \times (1 - (1 - e^{2\pi i (r_1 - r_2)/L}) u_j)^N u_j^{-1+2/\beta} (1 - u_j)^{-1+2/\beta} \prod_{j<k} |u_k - u_j|^{4/\beta}.
\end{aligned} \tag{1.51}$$

1.5. *Scaled limits*

As remarked in Section 1.1, the log-gas picture predicts the leading order form of the density profile for the Jacobi, Laguerre and Gaussian β-ensembles. It must then be that these same functional forms are the leading asymptotic form of the corresponding multidimensional integrals, appropriately scaled, in (1.38), (1.41), (1.46). Saddle point analysis undertaken in [16, 2] has verified that this is indeed the case. The analysis has been extended for the Gaussian and Laguerre β-ensembles in [7] to the calculation of correction terms. In particular, for the Gaussian β-ensemble it is found that

$$\begin{aligned}
\sqrt{\frac{2}{N}}\rho_{(1)}(\sqrt{2N}x) &\sim \rho_{\mathrm{W}}(x) - \frac{2}{\pi} \frac{\Gamma(1+2/\beta)}{(\pi\rho_{\mathrm{W}}(x))^{6/\beta-1}} \frac{1}{N^{2/\beta}} \\
&\quad \times \cos\Big(2\pi N P_{\mathrm{W}}(x) + (1-2/\beta)\mathrm{Arcsin}\, x\Big) + \mathrm{O}\Big(\min\Big(\frac{1}{N}, \frac{1}{N^{8/\beta}}\Big)\Big)
\end{aligned} \tag{1.52}$$

where $\rho_{\mathrm{W}}(x) := \frac{2}{\pi}\sqrt{1-x^2}$ and

$$P_{\mathrm{W}}(x) = \int_{-1}^{x} \rho_{\mathrm{W}}(t)\, dt = 1 + \frac{x}{2}\rho_{\mathrm{W}}(x) - \frac{1}{\pi}\mathrm{Arccos}\, x.$$

The expansion (1.52) is an example of a global asymptotic form, in which the expansion parameter varies macroscopically relative to the inter-particle spacing. In contrast local asymptotic expansions fix the inter-particle spacing to be order unity. In the circular β-ensemble defined on a circle of circumference length L as assumed in (1.47), this is achieved by taking the limit $N, L \to \infty$ with $N/L = \rho$ fixed. This limiting procedure applied to (1.49) gives [14]

$$\begin{aligned}\rho_{(n)}^{\text{bulk}}(r_1, \ldots, r_n) &:= \lim_{\substack{N,L\to\infty \\ N/L=\rho}} \rho_{(n)}(r_1, \ldots, r_n) \\ &= \rho^n c_n(\beta) \prod_{1\le j<k\le n} |2\pi\rho(r_k - r_j)|^\beta \prod_{k=2}^{n} e^{\pi i\rho\beta(r_k - r_1)} \\ &\quad \times {}_1F_1^{(\beta/2)}(n, 2n; -2\pi i\rho(r_2 - r_1), \ldots, -2\pi i\rho(r_n - r_1))\end{aligned}$$

where in the argument of ${}_1F_1^{(\beta/2)}$ each $-2\pi i\rho(r_j - r_1)$ $(j = 2, \ldots, n)$ occurs β times, and

$$c_n(\beta) = (\beta/2)^{\beta n(n-1)/2}((\beta/2)!)^n \prod_{k=0}^{n-1} \frac{\Gamma(\beta k/2 + 1)}{\Gamma(\beta(n+k)/2 + 1)}.$$

For the 2-point function, applying the limit to (1.51) gives [15]

$$\begin{aligned}\rho_{(2)}^{\text{bulk}}(r_1, r_2) &= \rho^2(\beta/2)^\beta \frac{((\beta/2)!)^3}{\beta!(3\beta/2)!} \frac{e^{-\pi i\beta\rho(r_1 - r_2)}(2\pi\rho(r_1 - r_2))^\beta}{S_\beta(-1 + 2/\beta, -1 + 2/\beta, 2/\beta)} \\ &\quad \times \int_{[0,1]^\beta} du_1 \cdots du_\beta \prod_{j=1}^{\beta} e^{2\pi i\rho(r_1 - r_2)u_j} u_j^{-1+2/\beta} \qquad (1.53) \\ &\quad \times (1 - u_j)^{-1+2/\beta} \prod_{j<k} |u_k - u_j|^{4/\beta}. \qquad (1.54)\end{aligned}$$

Local expansions can also be performed in the neighborhood of the edge of the support. There are two distinct cases: the spectrum edge when the support is strictly zero in one direction as near $x = 0$ in the Laguerre β-ensemble or at both edges of the support of the Jacobi β-ensemble, and the spectrum edge when the support is non-zero in both directions about the edge. These are referred to as the hard and soft edges respectively.

For the hard edge the necessary scaling is $x \mapsto X/4N$. We see from (1.40) and (1.26) that

$$\begin{aligned}\rho_{(n)}^{\rm hard}(X_1,\dots,X_n) &= \lim_{N\to\infty}\left(\frac{1}{4N}\right)^n \rho_{(n)}(X_1/4N,\dots,X_n/4N)\\ &= A_n(\beta)\prod_{j=1}^n X_j^{\beta a/2}\prod_{1\le j<k\le n}|X_k - X_j|^\beta\\ &\quad\times {}_0F_1^{(\beta/2)}(a+2n;Y_1,\dots,Y_{n\beta})\Big|_{\{Y_j\}\mapsto\{-X_j/4\}}\end{aligned} \tag{1.55}$$

where

$$\begin{aligned}A_n(\beta) &= 2^{-n(2+a\beta+\beta(n-1))}(\beta/2)^{n(1+a\beta+\beta(n-1))}\\ &\quad\times\frac{(\Gamma(1+\beta/2))^n}{\prod_{j=1}^{2n}\Gamma(1+a\beta/2+\beta(j-1)/2)}.\end{aligned}$$

With $n=1$, and $a = c - 2/\beta$, c a positive integer, the generalized hypergeometric function ${}_0F_1^{(\beta/2)}$ can be written as a β-dimensional integral to give [17]

$$\begin{aligned}\rho_{(1)}(X) &= a(c,\beta)X^{\beta/2-1}\int_{[-\pi,\pi]^\beta}\prod_{j=1}^\beta e^{iX^{1/2}\cos\theta_j}e^{i(c-1)\theta_j}\\ &\quad\times\prod_{1\le j<k\le\beta}|e^{i\theta_k}-e^{i\theta_j}|^{4/\beta}d\theta_1\cdots d\theta_\beta\end{aligned} \tag{1.56}$$

where

$$a(c,\beta) = (-1)^{(c-1)\beta/2}(2\pi)^{-\beta}\frac{1}{2}\left(\frac{\beta}{4}\right)^\beta\frac{(\Gamma(1+2/\beta))^\beta}{\Gamma(\beta)}.$$

Similarly the hard edge scaled limits can be taken in the evaluations of the distributions (1.43) and (1.45). Thus one finds [16]

$$\begin{aligned}E_\beta(0;(0,s)) &= e^{-\beta s/8}{}_0F_1^{(\beta/2)}(2m/\beta;x_1,\dots,x_m)\Big|_{x_j=-s/4}\\ p_\beta(0;s) &= A_{m,\beta}s^m e^{-\beta s/8}{}_0F_1^{(\beta/2)}(2m/\beta+2;;x_1,\dots,x_m)\Big|_{x_j=-s/4}\end{aligned}$$

where

$$A_{m,\beta} = 4^{-(m+1)}(\beta/2)^{2m+1}\frac{\Gamma(1+\beta/2)}{\Gamma(1+m)\Gamma(1+m+\beta/2)}.$$

Note the similarity with (1.55) in the case $n=1$. In particular we have available m-dimensional integral representations.

At the soft edge the appropriate scaling is $x \mapsto \sqrt{2N} + \frac{x}{\sqrt{2}N^{1/3}}$. Starting with the formula (1.46) one can show [7]

$$\frac{1}{\sqrt{2}N^{1/3}}\rho_{(1)}\Big(\sqrt{2N} + \frac{x}{\sqrt{2}N^{1/3}}\Big)$$
$$\sim \frac{\Gamma(1+\beta/2)}{2\pi}\Big(\frac{4\pi}{\beta}\Big)^{\beta/2}\prod_{j=1}^{\beta}\frac{\Gamma(1+2/\beta)}{\Gamma(1+2j/\beta)}K_{\beta,\beta}(x) + \mathrm{O}(N^{-1/3}) \quad (1.57)$$

where

$$K_{n,\beta}(x) := -\frac{1}{(2\pi i)^n}\int_{-i\infty}^{i\infty} dv_1 \cdots \int_{-i\infty}^{i\infty} dv_n \prod_{j=1}^{n} e^{v_j^3/3 - xv_j} \prod_{1\le k<l\le n} (v_k - v_j)^{4/\beta}$$

2. Physical Random Matrix Ensembles

2.1. *Heavy nuclei and quantum mechanics*

Random matrices were introduced into theoretical physics in the 1950's by Wigner as a model of a random matrix approximation to the Hamiltonian determining the highly excited states of heavy nuclei (see [30] for a collection of many early works in the field). At the time it was thought that the complex structure of heavy nuclei meant that in any basis the matrix elements for the Hamiltonian determining the highly excited states would effectively be random. (Subsequently [3] it has been learnt that a random matrix hypothesis applies equally well to certain single particle quantum systems; what is essential is that the underlying classical mechanics is chaotic.) One crucial point was the understanding that the (global) time reversal symmetry exhibited by complex nuclei implied that the elements of the matrix could be chosen to be symmetric, which since Hamiltonians are Hermitian implied the relevant class of matrices to be real symmetric. Another crucial point was the hypothesis of there being no preferential basis, in the sense that the joint probability distribution of the independent elements of the random matrix X should be independent of the basis vectors used to construct the matrix in the first place. This effectively requires that the joint probability distribution be unchanged upon the conjugation $X \mapsto O^T X O$ where O is a real orthogonal matrix. Distributions with this property are said to be orthogonally invariant, a typical example being

$$\frac{1}{C}e^{-\mathrm{Tr}(V(X))} \quad (2.1)$$

where C denotes the normalization, and $V(x)$ is a polynomial in x of even degree with positive leading coefficient. It is a well-known result that if one should require that the independent elements be independently distributed, and that the distribution be orthogonally invariant, then the distribution is necessarily of the form (2.1) with

$$V(x) = ax^2 + bx. \tag{2.2}$$

Random matrix theory applied to quantum systems without time reversal symmetry (typically due to the presence of a magnetic field) gives the relevant class of matrices as being complex Hermitian. In this case the hypothesis of there being no preferential basis requires invariance of the joint probability distribution under the conjugation $X \mapsto U^\dagger X U$ where U is a unitary matrix. Distributions with this property are said to be unitary invariant, and again (2.1) is a typical example, with X now a complex Hermitian rather than real symmetric matrix.

Theoretically a time reversal operator T is any anti-unitary operator. However physical considerations further restricts their form (see e.g. [23]), requiring that for an even number or no spin 1/2 particles $T^2 = 1$ (a familiar example being $T = K$ where K denotes complex conjugation), while for a finite dimensional system with an odd number of spin 1/2 particles $T^2 = -1$ where $T = \mathbb{Z}_{2N} K$, with

$$\mathbb{Z}_{2N} = 1_N \otimes \begin{bmatrix} 0 & -1 \\ 1 & 0 \end{bmatrix}.$$

For a quantum system which commutes with T of this latter form, the $2N \times 2N$ matrix X modelling the Hamiltonian must, in addition to being Hermitian, have the property

$$X = \mathbb{Z}_{2N} \bar{X} \mathbb{Z}_{2N}^{-1}.$$

This means that X can be viewed as the $2N \times 2N$ complex matrix formed from an $N \times N$ matrix with the elements consisting of 2×2 blocks of the form

$$\begin{bmatrix} 0 & -1 \\ 1 & 0 \end{bmatrix}, \tag{2.3}$$

which is the matrix representation of a real quaternion. For no preferential basis one requires that the probability density function for X be invariant under the conjugation $X \mapsto S^\dagger X S$ where S is the $2N \times 2N$ unitary matrix formed out of an $N \times N$ unitary matrix with real quaternian blocks (2.3).

2.2. *Dirac operators and QCD*

Random Hermitian matrices with a special block structure occurred in Verbaarschot's introduction [36] of a random matrix theory of massless Dirac operators, in the context of quantum chromodynamics (QCD). Generally the non-zero eigenvalues of the massless Dirac operator occur in pairs $\pm\lambda$. Furthermore, in the so called chiral basis, all basis elements are eigenfunctions of the γ-matrix $i\gamma_5$ with eigenvalues $+1$ or -1, and matrix elements between states with the same eigenvalue of γ_5 must vanish, leaving a block structure with non-zero elements in the upper-right and lower-left blocks only. Noting too that the application to QCD requires that the Dirac operator has a given number, ν say, of zero eigenvalues, implies the structure

$$\begin{bmatrix} 0_{n\times n} & X \\ X^\dagger & 0_{m\times m} \end{bmatrix} \tag{2.4}$$

where X is an $n \times m$ $(n \geq m)$ matrix with $n - m = \nu$. Moreover the positive eigenvalues are given by the positive square root of the eigenvalues of $X^\dagger X$.

As in the application to chaotic quantum systems, the elements of X in (2.4) must be real, complex or real quaternion according to there being a time reversal symmetry with $T^2 = 1$, no time reversal symmetry, or a time reversal symmetry with $T^2 = -1$ respectively. And due to their origin in studying the Dirac equation with a chiral basis, the corresponding ensembles are referred to as chiral random matrices.

2.3. *Random scattering matrices*

Problems in quantum physics also give rise to random unitary matrices. One such problem is the scattering of plane waves within an irregular shaped domain, or one containing random scattering impurities. The wave guide connecting to the cavity is assumed to permit N distinct plane wave states, and the corresponding amplitudes are denoted $\vec{I}$ for the ingoing states and $\vec{O}$ for the outgoing states. By definition the scattering matrix S relates $\vec{I}$ and $\vec{O}$,

$$S\vec{I} = \vec{O}.$$

The flux conservation requirement $|\vec{I}|^2 = |\vec{O}|^2$ implies that S must be unitary. For scattering matrices in quantum mechanics, or more generally evolution operators, time reversal symmetry requires that

$$T^{-1}ST = S^\dagger.$$

For $T^2 = 1$ this implies

$$S = S^T \tag{2.5}$$

while for $T^2 = -1$,

$$S = \mathbb{Z}_{2N} S^T \mathbb{Z}_{2N}^{-1} =: S^D. \tag{2.6}$$

With $U_N \in U(N)$ a general $N \times N$ unitary matrix note that

$$S = U_N U_N^T, \qquad S = U_{2N} U_{2N}^D \tag{2.7}$$

respectively have the properties (2.5) and (2.6).

For random scattering matrices it is hypothesized that the statistical properties of S are determined soley by the global time reversal symmetry, and are invariant under the same conjugations as the corresponding Hamiltonians. In fact the measure on $U(N)$ which is unchanged by both left and right multiplication by another unitary matrix is unique. It is called the Haar measure $d_H U$, and its volume form is

$$(d_H U) = (U^\dagger dU). \tag{2.8}$$

Similarly, for the symmetric and self dual quaternion unitary matrices in (2.8) we have

$$(d_H S) = ((U_N^T)^\dagger dS\, U^\dagger) \qquad (d_H S) = ((U_N^T)^\dagger dS\, U^\dagger) \tag{2.9}$$

which are invariant under the mappings

$$S \mapsto V_N^T S V_N, \qquad S \mapsto V_{2N}^D S V_{2N}$$

for general unitary matrices V.

2.4. *Quantum conductance problems*

In the above problem of scattering within a cavity the incoming and outgoing wave is unchanged along the lead. A related setting is a quasi one-dimensional conductor (lead) which contains internal scattering impurities (see e.g. [34]). One supposes that there are n available scattering channels at the left hand edge, m at the right hand edge, and that at each end there is a reservoir which causes a current to flow.

The n-component vector $\vec{I}$ and m-component vector $\vec{I'}$ is used to denote the amplitudes of the plane wave states traveling into the left and right sides of the wire respectively, while the n-component vector $\vec{O}$ and m-component vector $\vec{O}'$ denotes the amplitudes of the plane wave states traveling out of the left and right sides of the wire. The $(n+m)\times(n+m)$ scattering matrix S now relates the flux traveling into the conductor to the flux traveling out,

$$S\begin{bmatrix}\vec{I}\\ \vec{I}'\end{bmatrix}=\begin{bmatrix}\vec{O}\\ \vec{O}'\end{bmatrix}.$$

The scattering matrix is further decomposed in terms of reflection and transmission matrices by

$$S=\begin{bmatrix}r_{n\times n} & t'_{n\times m}\\ t_{m\times n} & r'_{m\times m}\end{bmatrix}. \tag{2.10}$$

According to the Landauer scattering theory of electronic conduction, the conductance G is given in terms of the transmission matrix $t_{m\times n}$ by

$$G/G_0=\mathrm{Tr}(t^\dagger t)$$

where $G_0=2e^2/h$ is twice the fundamental quantum unit of conductance. Thus of interest is the distribution of $t^\dagger t$ in the case S is a random unitary matrix (no time reversal symmetry), a symmetric unitary matrix (time reversal symmetry $T^2=1$), or has the self dual property (2.6) (time reversal symmetry $T^2=-1$).

2.5. *Eigenvalue p.d.f.'s for Hermitian matrices*

Let $X=[x_{jk}]_{j,k=1,\dots,N}$ be a real symmetric matrix. The diagonal and upper triangular entries are independent variables. We know that real symmetric matrices can be diagonalized according to

$$X=OLO^T$$

where $L=\mathrm{diag}(\lambda_1,\dots,\lambda_N)$ is the diagonal matrix of the eigenvalues (a total of N independent variables), while O is a real orthogonal matrix formed out of the eigenvectors (a total of $N(N-1)/2$ independent variables). We seek the Jacobian for the change of variables from the independent elements of X, to the eigenvalues $\lambda_1,\dots,\lambda_N$ and the $N(N-1)/2$ linearly independent variables formed out of linear combinations of the elements of O.

To calculate the Jacobian, it is useful to be familiar with the notion of the wedge product,

$$du_1\wedge\cdots\wedge du_n:=\det[du_i(\vec{r}_j)]_{i,j=1,\dots,n}. \tag{2.11}$$

Now, when changing variables from $\{u_1, \dots, u_n\}$ to $\{v_1, \dots, v_n\}$, since

$$du_i = \sum_{l=1}^{n} \frac{\partial u_i}{\partial v_l} dv_l$$

and

$$\left[\sum_{l=1}^{n} \frac{\partial u_i}{\partial v_l} dv_l(\vec{r}_j) \right]_{i,j=1,\dots,n} = \left[\frac{\partial u_i}{\partial v_j} \right]_{i,j=1,\dots,n} [dv_i(\vec{r}_j)]_{i,j=1,\dots,n}$$

it follows from (2.11) that

$$du_1 \wedge \cdots \wedge du_n = \det \left[\frac{\partial u_i}{\partial v_j} \right]_{i,j=1,\dots,n} dv_1 \wedge \cdots \wedge dv_n$$

thus allowing the Jacobian to be read off.

Denote by dH denote the matrix of differentials of H. We have

$$dH = dO\, LO^T + OdL\, O^T + OLdO^T.$$

Noting from $O^T O = 1_N$ that $dO^T\, O = -O^T dO$ it follows from this that

$$\begin{aligned} O^T dH\, O &= O^T dO\, L - LO^T dO + dL \\ &= \begin{bmatrix} d\lambda_1 & (\lambda_2 - \lambda_1)\vec{o}_1^{\,T} \cdot d\vec{o}_2 & \cdots & (\lambda_N - \lambda_1)\vec{o}_1^{\,T} \cdot d\vec{o}_N \\ (\lambda_2 - \lambda_1)\vec{o}_1^{\,T} \cdot d\vec{o}_2 & d\lambda_2 & \cdots & (\lambda_N - \lambda_2)\vec{o}_2^{\,T} \cdot d\vec{o}_N \\ \vdots & \vdots & & \vdots \\ (\lambda_N - \lambda_1)\vec{o}_1^{\,T} \cdot d\vec{o}_N & (\lambda_N - \lambda_2)\vec{o}_2^{\,T} \cdot d\vec{o}_N & \cdots & d\lambda_N \end{bmatrix} \end{aligned}$$

where $\vec{o}_k$ denotes the kth column of O.

For H Hermitian, let (dH) denote (up to a sign) the wedge product of all the independent elements, real and imaginary parts separately, of H. To compute the wedge product on the left hand side, the following result is required (see e.g. [13]).

Proposition 2.1. *Let A and M be $N \times N$ matrices, where A is non-singular. For A real ($\beta = 1$), complex ($\beta = 2$) or real quaternion ($\beta = 4$), and M real symmetric ($\beta = 1$), complex Hermitian ($\beta = 2$) or quaternion real Hermitian ($\beta = 4$)*

$$(A^\dagger dM\, A) = \left(\det A^\dagger A \right)^{\beta(N-1)/2+1} (dM).$$

Making use of this result with $\beta = 1$ we see immediately that

$$(dH) = \prod_{1 \le j < k \le N} (\lambda_k - \lambda_j) \bigwedge_{j=1}^{N} d\lambda_j (O^T dO). \tag{2.12}$$

A simple scaling argument can be used to predict the structure of (2.12). Since there are $N(N+1)/2$ independent differentials in (dH), we see that for c a scalar

$$(dcH) = c^{N(N+1)/2}(dH).$$

But $cH = OcLO^T$, so we conclude that (dcH) is a homogeneous polynomial of degree $N(N-1)/2$ in $\{\lambda_j\}$ (note that $d\lambda_1 \cdots d\lambda_N$ contributes degree N). Furthermore, because the probability of repeated eigenvalues occurs with zero probability (dX) must vanish for $\lambda_j = \lambda_k$. These two facts together tell us that the dependence in the eigenvalues is precisely as in (2.12).

The analogue of (2.12) for complex Hermitian matrices is

$$(dH) = \prod_{1 \le j < k \le N} (\lambda_k - \lambda_j)^2 \bigwedge_{j=1}^{N} d\lambda_j (U^\dagger dU) \tag{2.13}$$

while for Hermitian matrices with real quaternion entries it is

$$(dH) = \prod_{1 \le j < k \le N} (\lambda_k - \lambda_j)^4 \bigwedge_{j=1}^{N} d\lambda_j (S^\dagger dS). \tag{2.14}$$

It is of interest to understand (2.13) and (2.14) from the viewpoint of scaling. Consider for definiteness (2.13) (the case of (2.14) is similar). Recalling that for H complex Hermitian (dH) consists of the product of differentials of the independent real and imaginary parts, it follows that $(dcH) = c^{N^2}(dH)$. This tells us that the polynomial in $\{\lambda_j\}$ in (2.13) is of degree $N^2 - N$, and we know from the argument in the case of (2.12) that it contains a factor of $\prod_{j<k}(\lambda_k - \lambda_j)$. We want to deduce that in fact this factor is repeated twice. For this we use the fact that the $N \times N$ complex Hermitian matrix $[x_{jk} + iy_{jk}]_{j,k=1,\dots,N}$ has the same eigenvalues as the $2N \times 2N$ real matrix

$$\begin{bmatrix} x_{jk} & y_{jk} \\ -y_{jk} & x_{jk} \end{bmatrix}_{j,k=1,\dots,N} \tag{2.15}$$

but with each eigenvalue doubly degenerate, due to the isomorphism between the complex numbers and the 2×2 matrices exhibited in (2.15). From this viewpoint the second factor then corresponds to a double degeneracy.

As a consequence of (2.12), (2.13), (2.14) it follows that the eigenvalue p.d.f. for ensembles of Hermitian matrices weighted by (2.1) and with real $(\beta = 1)$, complex $(\beta = 2)$ and real quaternion $(\beta = 4)$ elements is

$$\frac{1}{C} e^{-\sum_{j=1}^{N} V(\lambda_j)} \prod_{1 \le j < k \le N} |\lambda_k - \lambda_j|^\beta.$$

2.6. *Eigenvalue p.d.f.'s for Wishart matrices*

For X a random $n \times m$ rectangular matrix ($n \geq m$), the positive definite matrix $A := X^\dagger X$ is referred to as a Wishart matrix, after the application of such matrices in multi-variate statistics (see e.g. [28]). In the latter setting there are m variables $x_1, \dots, x_m$, each measured n times, to give an array of data $[x_k^{(j)}]_{\substack{j=1,\dots,n\\k=1,\dots,m}}$ which is thus naturally represented as a matrix. Then

$$X^T X = \left[\sum_{j=1}^n x_{k_1}^{(j)} x_{k_2}^{(j)} \right]_{k_1,k_2=1,\dots,m}$$

is essentially an empirical approximation to the covariance matrix for the data.

Fundamental to the computation of the eigenvalue p.d.f. for Wishart matrices is the following result relating the Jacobian for changing variables from the elements of X to the elements of A (and other associated variables not explicitly stated).

Proposition 2.2. *Let the $n \times m$ matrix X have real ($\beta = 1$), complex ($\beta = 2$) or real quaternion ($\beta = 4$) elements, and suppose it has p.d.f. of the form $F(X^\dagger X)$. The p.d.f. of $A := X^\dagger X$ is then proportional to*

$$F(A)\Big(\det A\Big)^{(\beta/2)(n-m+1-2/\beta)}.$$

Proof. We follow [29]. The p.d.f. of A must be equal to

$$F(A)h(A)$$

for some h. Write $A = B^\dagger V B$ where V is positive definite. Making use of the result of Proposition 2.1 tells us that the p.d.f. of V is then

$$F(B^\dagger VB)h(B^\dagger VB)\det(B^\dagger B)^{(\beta/2)(m-1+2/\beta)}. \tag{2.16}$$

Now let $X = YB$, where Y is such that $V = Y^\dagger Y$. Noting that for $\vec{x}^{\,T} = \vec{y}^{\,T} B$, the Jacobian is $(\det B^\dagger B)^{\beta/2}$, it follows that

$$(dX) = (\det B^\dagger B)^{\beta n/2}(dY)$$

and hence the p.d.f. of Y is

$$F(B^\dagger Y^\dagger YB)(\det B^\dagger B)^{\beta n/2}.$$

This is a function of $Y^\dagger Y$, so the p.d.f. of $V = Y^\dagger Y$ is

$$F(B^\dagger VB)(\det B^\dagger B)^{\beta n/2}h(V). \tag{2.17}$$

Equating (2.16) and (2.17) gives

$$h(B^\dagger V B) = h(V)(\det B^\dagger B)^{(\beta/2)(n-m+1-2/\beta)}.$$

Setting $V = 1_m$ and noting $h(1_m) = c$ for some constant c implies the sought result. □

Suppose that, analogous to (2.1), the matrix X in (2.4) is distributed according to

$$\frac{1}{C} e^{-\mathrm{Tr}(V(X^\dagger X))}$$

where $V(x)$ is a polynomial in x with positive leading term. Then as a consequence of Proposition 2.2 and (2.12)–(2.14) the eigenvalue p.d.f. of $A = X^\dagger X$ is equal to

$$\frac{1}{C} e^{-\sum_{j=1}^m V(\lambda_j)} \prod_{j=1}^m \lambda_j^{(\beta/2)(n-m+1-2/\beta)} \prod_{1 \le j < k \le m} |\lambda_k - \lambda_j|^\beta \tag{2.18}$$

where $0 \le \lambda_j < \infty$ $(j = 1, \ldots, N)$. Because the eigenvalues of (2.4), $\{x_j\}$ say, are related to the eigenvalues $\{\lambda_j\}$ by $x_j^2 = \lambda_j$, it follows that the $\{x_j\}$ have p.d.f.

$$\frac{1}{C} e^{-\sum_{j=1}^m V(x_j^2)} \prod_{j=1}^m |x_j|^{\beta(n-m+1-2/\beta)+1} \prod_{1 \le j < k \le m} |x_k^2 - x_j^2|^\beta. \tag{2.19}$$

2.7. *Eigenvalue p.d.f.'s for unitary matrices*

We now take up the problem of computing the Haar volume form (2.8) and its analogues (2.9) for symmetric and self dual quaternion unitary matrices. There are a number of possible approaches (see e.g. [13]). Here use will be made of the Cayley transform

$$H = i\frac{1_N - U}{1_N + U} \tag{2.20}$$

which maps a unitary matrix U to an Hermitian matrix H. From this the volume form $(U^\dagger dU)$ can be computed in terms of (dH), and the decomposition of the latter in terms of its eigenvalues and eigenvectors is already known. To begin we invert (2.20) so it reads

$$U = \frac{1_N + iH}{1_N - iH}.$$

Making use of the general operator identity

$$\frac{d}{da}(1-K)^{-1} = (1-K)^{-1}\frac{dK}{da}(1-K)^{-1},$$

where K is assumed to be a smooth function of a, we deduce from this that

$$U^\dagger dU = 2i(1_N + iH)^{-1} dH (1_N - iH)^{-1}. \tag{2.21}$$

To be consistent with (2.7), (2.9) for U symmetric or self dual quaternion we introduce the decompositions

$$U = V^T V, \qquad U = V^D V$$

for $V \in U(N)$, $V \in U(2N)$ respectively, and use (2.21) to calculate $\delta U := VU^\dagger dU\, V^\dagger$. This gives

$$\delta U = \frac{i}{2}(V^\phi + V^\dagger)^\phi dH (V^\phi + V^\dagger)$$

where $\phi = T$ $(\beta = 1)$, $\phi = D$ $(\beta = 4)$. Observe that the elements of $V^\phi + V^\dagger$ are real for $\phi = T$, real quaternion for $\phi = D$. This tells us that Proposition 2.1 can be applied to the right hand side of (2.21) with appropriate β and thus

$$(\delta U) = 2^{N(\beta(N-1)/2+1)}\Big(\det(1_N + H^2)\Big)^{-\beta(N-1)/2-1}(dH). \tag{2.22}$$

Let $\{\lambda_j\}$ denote the eigenvalues of H and $\{e^{i\theta_j}\}$ denote the corresponding eigenvalues of U, with H and U related by (2.20). Then

$$e^{i\theta} = \frac{1+i\lambda}{1-i\lambda}. \tag{2.23}$$

From (2.22) the corresponding eigenvalues p.d.f. of $\{\lambda_j\}$ is

$$\frac{1}{C}\prod_{l=1}^N \frac{1}{(1+\lambda_l^2)^{\beta(N-1)/2+1}} \prod_{j<k} |\lambda_k - \lambda_j|^\beta. \tag{2.24}$$

Changing variables according to (2.23) gives for the eigenvalue p.d.f. of $\{e^{i\theta_j}\}$

$$\frac{1}{C}\prod_{j<k} |e^{i\theta_k} - e^{i\theta_j}|^\beta. \tag{2.25}$$

2.8. *Eigenvalue p.d.f.'s for blocks of unitary matrices*

We seek the distribution of the non-zero eigenvalues of $t^\dagger t$ in the decomposition (2.10). To compute this distribution, one approach is to consider the singular value decomposition of each of the individual blocks, for example $t_{m\times n} = U_t \Lambda_t V_t^\dagger$, where Λ_t is a rectangular diagonal matrix with diagonal entries consisting of the square roots of the non-zero eigenvalues of $t^\dagger t$, and U_t and V_t are $m \times m$ and $n \times n$ unitary matrices. In terms of such decompositions it is possible to parametrize (2.10) as

$$S = \begin{bmatrix} U_r & 0 \\ 0 & U_{r'} \end{bmatrix} L \begin{bmatrix} V_r^\dagger & 0 \\ 0 & V_{r'}^\dagger \end{bmatrix} \tag{2.26}$$

where

$$L = \begin{bmatrix} \sqrt{1 - \Lambda_t \Lambda_t^T} & i\Lambda_t \\ i\Lambda_t^T & \sqrt{1 - \Lambda_t^T \Lambda_t} \end{bmatrix}.$$

In the case that S is symmetric, it is further required that

$$V_r^\dagger = U_r^T, \qquad V_{r'}^\dagger = U_{r'}^T, \tag{2.27}$$

while for S self dual quaternion

$$V_r^\dagger = U_r^D, \qquad V_{r'}^\dagger = U_{r'}^D. \tag{2.28}$$

From (2.26) the method of wedge products can be used to derive that the non-zero elements of Λ_t have the distribution

$$\prod_{j=1}^m \lambda_j^{\beta\alpha} \prod_{1\le j<k\le m} |\lambda_k^2 - \lambda_j^2|^\beta, \qquad \alpha = n - m + 1 - 2/\beta \tag{2.29}$$

where $0 < \lambda_j < 1$ $(j = 1, \dots, m)$. But it turns out that the details of the calculation are quite tedious [13, 19]. In the case $\beta = 2$ some alternative derivations are possible [33, 6, 19], and a more general result can be derived.

Proposition 2.3. *Let U be an $N \times N$ random unitary matrix chosen with Haar measure. Decompose U into blocks*

$$U = \begin{bmatrix} A_{n_1 \times n_2} & C_{n_1 \times (N-n_2)} \\ B_{(N-n_1)\times n_2} & D_{(N-n_1)\times(N-n_2)} \end{bmatrix} \tag{2.30}$$

where $n_1 \ge n_2$. The eigenvalue p.d.f. of $Y := A^\dagger A$ is proportional to

$$\prod_{j=1}^{n_2} y_j^{(n_1-n_2)} (1 - y_j)^{(N-n_1-n_2)} \prod_{j<k} (y_k - y_j)^2. \tag{2.31}$$

Proof. We will use the matrix integral [22]

$$\int e^{(i/2)\mathrm{Tr}(HQ)} \Big(\det(H - \mu 1_m) \Big)^{-n} (dH) \propto (\det Q)^{(n-m)} e^{(i/2)\mu \mathrm{Tr} Q}, \quad (2.32)$$

valid for Q Hermitian and $\mathrm{Im}(\mu) > 0$, and the integration is over the space of $m \times m$ Hermitian matrices. In (2.30) the fact that U is unitary tells us that

$$AA^\dagger + CC^\dagger = 1_{n_1}. \quad (2.33)$$

Following an idea of [38], we regard (2.33) as a constraint in the space of general $n_1 \times n_2$ and $n_1 \times (N - n_2)$ complex rectangular matrices A and C, which allows the distribution of A to be given by

$$\int \delta(AA^\dagger + CC^\dagger - 1_{n_2})(dC). \quad (2.34)$$

The delta function in (2.34) is a product of scalar delta functions, which in turn is proportional to the matrix integral

$$\int e^{-i\mathrm{Tr}(H(AA^\dagger + CC^\dagger - 1_{n_2}))}(dH), \quad (2.35)$$

where the integration is over the space of $n_2 \times n_2$ Hermitian matrices.

Substituting (2.35) and (2.34) and changing the order of integration, the integration over C is a Gaussian integral and so can be computed immediately. However for the resulting function of H to be integrable around $H = 0$, the replacement $H \mapsto H - i\mu I_n$ in the exponent of (2.35) must be made. Doing this we are able to deduce (2.34) to be proportional to

$$\lim_{\mu \to 0+} \int (\det(H - i\mu 1_{n_1}))^{-(N-n_2)} e^{i\mathrm{Tr}(H(1_{n_1} - AA^\dagger))}(dH) \quad (2.36)$$

which in turn is proportional to

$$(\det(1_{n_1} - AA^\dagger))^{(N-n_1-n_2)} \quad (2.37)$$

where (2.37) follows from (2.36) using (2.32). Because the non-zero eigenvalues of $AA^\dagger$ and $A^\dagger A$ are the same, we can replace $AA^\dagger \mapsto A^\dagger A$ in (2.37). Now using Proposition 2.2 in the case $\beta = 2$ gives that the distribution of A is proportional to

$$(\det Y)^{(n_1-n_2)}(\det(1_{n_2} - Y))^{(N-n_1-n_2)}. \quad (2.38)$$

Changing variables now to the eigenvalues and eigenvectors using (2.13) gives the stated result. □

The eigenvalue p.d.f. (2.31) reclaims (2.30) in the case $\beta = 2$ by setting $n_1 = n_2 = m$, $N = n + m$ and changing variables $1 - y_j = \lambda_j^2$.

2.9. *Classical random matrix ensembles*

Let $\beta = 1, 2$ or 4 according to the elements being real, complex or quaternion real respectively. In the case of Hermitian matrices, the eigenvalue p.d.f.'s derived above all have the general form

$$\frac{1}{C}\prod_{l=1}^{N} g(x_l) \prod_{1\le j<k\le N} |x_k - x_j|^{\beta}.$$

Choosing the entries of the matrices to be independent Gaussians, when there is a choice, the form of $g(x)$ is, up to scaling $x_l \mapsto cx_l$,

$$g(x) = \begin{cases} e^{-x^2}, & \text{Gaussian} \\ x^a e^{-x} \ (x>0) & \text{Laguerre} \\ x^a(1-x)^b \ (0<x<1) & \text{Jacobi} \\ (1+x^2)^{-\alpha} & \text{Cauchy.} \end{cases}$$

These are the four classical weight functions from orthogonal polynomial theory, which can be characterized by the property that

$$\frac{d}{dx}\log g(x) = \frac{a(x)}{b(x)}$$

where

$$\operatorname{degree} a(x) \le 1, \qquad \operatorname{degree} b(x) \le 2.$$

Recall that stereographic projection of the Cauchy weight for a certain α gives the circular ensemble, as noted in (2.23)–(2.25). Thus these are essentially the same p.d.f.'s encountered in the log-gas systems of Section 1.1, and the quantum many body systems of Section 1.2, except that β is restricted to one of three values.

It is our objective in the rest of these notes to explore some eigenvalue problems which relate to the Gaussian and Laguerre β ensembles for general $\beta > 0$.

3. β-Ensembles of Random Matrices

3.1. *Gaussian β ensemble*

We will base our construction on an inductive procedure. Let a be a scalar chosen from a particular probability distribution, and let $\vec{w}$ be a $N \times 1$ column vector with each component drawn from a particular probability density. Inductively define a sequence of matrices $\{M_j\}_{j=1,2,\ldots}$ by $M_1 = a$ and

$$M_{N+1} = \begin{bmatrix} D_N & \vec{w} \\ \vec{w}^T & a \end{bmatrix} \tag{3.1}$$

where $D_N = \operatorname{diag}(a_1, \ldots, a_N)$ with $\{a_j\}$ denoting the eigenvalues of M_N. For example, suppose $a \in \mathrm{N}[0,1]$ and $w_j \in \mathrm{N}[0, 1/\sqrt{2}]$ $(j = 1, \ldots, N)$. Let O_N be the real orthogonal matrix which diagonalizes M_N, so that $M_N = O_N D_N O_N^T$, and observe

$$\begin{bmatrix} O_N & \vec{0} \\ \vec{0}^T & 1 \end{bmatrix} \begin{bmatrix} M_N & \vec{w} \\ \vec{w}^T & a \end{bmatrix} \begin{bmatrix} O_N & \vec{0} \\ \vec{0}^T & 1 \end{bmatrix}^T \sim \begin{bmatrix} D_N & \vec{w} \\ \vec{w}^T & a \end{bmatrix}.$$

It follows that the construction (3.1) gives real symmetric matrices M_N with distribution proportional to

$$e^{-\mathrm{Tr}(M_N^2)/2}$$

and we know the corresponding eigenvalue p.d.f. is

$$\frac{1}{C} \prod_{l=1}^{N} e^{-a_j^2/2} \prod_{1 \le j < k \le N} |a_k - a_j|. \tag{3.2}$$

Given the eigenvalues $\{a_j\}_{j=1,\ldots,N}$ of M_N we would like to compute the eigenvalues $\{\lambda_j\}_{j=1,\ldots,N+1}$ of M_{N+1}. Now

$$\begin{aligned} \det(\lambda 1_{N+1} - M_{N+1}) &= \det \begin{bmatrix} \lambda 1_N - D_N & -\vec{w} \\ -\vec{w}^T & \lambda - a \end{bmatrix} \\ &= \det \begin{bmatrix} \lambda 1_N - D_N & -\vec{w} \\ \vec{0}^T & \lambda - a - \vec{w}^T (\lambda 1_N - D_N)^{-1} \vec{w} \end{bmatrix} \\ &= p_N(\lambda)(\lambda - a - \vec{w}^T (\lambda 1_N - D_N)^{-1} \vec{w}) \end{aligned}$$

where $p_N(\lambda)$ is the characteristic polynomial for M_N. But $\lambda 1_N - D_N$ is diagonal, so its inverse is also diagonal, allowing us to conclude

$$\frac{p_{N+1}(\lambda)}{p_N(\lambda)} = \lambda - a - \sum_{i=1}^{N} \frac{q_i}{\lambda - a_i}, \qquad q_i := w_i^2. \tag{3.3}$$

The eigenvalues of M_{N+1} are thus given by the zeros of the rational function in (3.3). The corresponding p.d.f. can be computed for a certain choice of the distribution of the q_i [12, 20].

Proposition 3.1. *Let $w_i^2 \sim \Gamma[\beta/2, 1]$ where $\Gamma[s, \sigma]$ refers to the gamma distribution, specified by the p.d.f. $\sigma^{-s} x^{s-1} e^{-x/\sigma}/\Gamma(s)$ $(x > 0)$. Given*

$$a_1 > a_2 > \cdots > a_N$$

the p.d.f. for the zeros of the random rational function

$$\lambda - a - \sum_{i=1}^{N} \frac{q_i}{\lambda - a_i}$$

is equal to

$$\frac{e^{a^2/2}}{(\Gamma(\beta/2))^N} \frac{\prod_{1 \le j < k \le N+1} (\lambda_j - \lambda_k)}{\prod_{1 \le j < k \le N} (a_j - a_k)^{\beta - 1}} \prod_{j=1}^{N+1} \prod_{p=1}^{N} |\lambda_j - a_p|^{\beta/2 - 1}$$
$$\times \exp\left(-\frac{1}{2} \left(\sum_{j=1}^{N+1} \lambda_j^2 - \sum_{j=1}^{N} a_j^2 \right) \right) \tag{3.4}$$

where

$$\infty > \lambda_1 > a_1 > \lambda_2 > \cdots > a_N > \lambda_{N+1} > -\infty \tag{3.5}$$

and

$$\sum_{j=1}^{N+1} \lambda_j = \sum_{j=1}^{N} a_j + a. \tag{3.6}$$

Proof. Because the q_i are positive, graphical considerations imply the interlacing condition. Note too that the summation constraint is equivalent to the statement that $\mathrm{Tr}\, M_{N+1} = \mathrm{Tr}\, D_N + a$, while the translations $\lambda_j \mapsto \lambda_j - a$, $a_j \mapsto a_j - a$ shows it suffices to consider the case $a = 0$.

With $a = 0$ we have

$$\lambda - \sum_{i=1}^{N} \frac{q_i}{\lambda - a_i} = \frac{\prod_{j=1}^{N+1} (\lambda - \lambda_j)}{\prod_{l=1}^{N} (\lambda - a_l)}.$$

From the residue at $\lambda = a_i$ it follows

$$\frac{\prod_{j=1}^{N+1} (a_i - \lambda_j)}{\prod_{l=1, l \ne i}^{N} (a_i - a_l)} = -q_i. \tag{3.7}$$

We want to compute the change of variables from $\{q_i\}_{i=1,\dots,N}$ to $\{\lambda_j\}_{j=1,\dots,N}$. It follows immediately from (3.7) that up to a sign

$$\bigwedge_{j=1}^{N} dq_i = \prod_{j=1}^{N} q_j \det\left[\frac{1}{a_i - \lambda_j}\right] \bigwedge_{j=1}^{N} d\lambda_j. \tag{3.8}$$

Hence after making use of the Cauchy double alternant identity the sought Jacobian is seen to be equal to

$$\prod_{j=1}^{N} q_j \left| \frac{\prod_{1\le i<j\le N} (a_i - a_j)(\lambda_i - \lambda_j)}{\prod_{i,j=1}^{N} (a_i - \lambda_j)} \right|. \tag{3.9}$$

But the distribution of $\{w_j\}$ is equal to

$$\frac{1}{(\Gamma(\beta/2))^N} \prod_{j=1}^{N} q_j^{\beta/2-1} \, e^{-\sum_{j=1}^{N} q_j}. \tag{3.10}$$

We must multiply (3.9) and (3.10), and write $\{q_j\}$ in terms of $\{a_i, \lambda_j\}$. By equating the coefficients of $1/\lambda$ on both sides of (3.3) and using (3.6) with $a = 0$ we see

$$\sum_{j=1}^{N} q_j = \frac{1}{2}\left(\sum_{j=1}^{N+1} \lambda_j^2 - \sum_{j=1}^{N} \mu_j^2 \right).$$

Further, we can read off $\prod_{i=1}^{N} q_i$ from (3.7). Substituting we deduce (3.4). □

Suppose now that

$$a \sim \mathrm{N}[0,1]. \tag{3.11}$$

Then we see from Proposition 3.1 that the (conditional) eigenvalue p.d.f. of $\{\lambda_j\}$ is given by (3.4) with $e^{a^2/2}$ replaced by the constant $\frac{1}{\sqrt{2\pi}}$, and the constraint (3.6) no longer present. Let this conditional eigenvalue p.d.f. be denoted $G_N(\{\lambda_j\},\{a_k\})$, and denote its domain of support (3.5) by R_N. Let $\{a_j\}$ have p.d.f. $p_N(a_1,\dots,a_N)$, and let $\{\lambda_j\}$ have p.d.f. $p_{N+1}(\lambda_1,\dots,\lambda_{N+1})$. With this notation, we read off from (3.4) that

$$p_{N+1}(\lambda_1,\dots,\lambda_{N+1}) = \int_R da_1 \cdots da_N \, G_N((\{\lambda_j\},\{a_k\}) p_N(a_1,\dots,a_N). \tag{3.12}$$

We seek the solution of this recurrence with $p_0 = 1$.

When $\beta = 1$ we know that the solution of (3.12) is given by (3.2). To obtain its solution for general β, we begin by noting that with μ_i denoting the top entry of the normalized eigenvector corresponding to the eigenvalue λ_i of M_N we have

$$\sum_{j=1}^{N} \frac{\mu_j^2}{\lambda - \lambda_j} = \Big((\lambda 1_N - M_N)^{-1} \Big)_{11} = \frac{p_{N-1}(\lambda)}{p_N(\lambda)}. \tag{3.13}$$

Here the first equality follows from the spectral decomposition, while the second follows from Cramer's rule. Because the matrix (3.1) is real symmetric and thus orthogonally diagonalizable, we must have

$$\sum_{j=1}^{N} \mu_i^2 = 1$$

which is consistent with (3.13).

In the case $\beta = 1$ the matrix M_N is orthogonally invariant and so we have

$$\mu_i^2 \sim \frac{w_i^2}{w_1^2 + \cdots + w_N^2} =: \rho_i$$

where each $w_i^2 \sim \Gamma[1/2, 1]$. Generally, if

$$w_i^2 \sim \Gamma[\beta/2, 1] \tag{3.14}$$

then the p.d.f. of $\rho_1, \ldots, \rho_N$ is equal to the Dirichlet distribution

$$\frac{\Gamma(N\beta/2)}{(\Gamma(\beta/2))^N} \prod_{j=1}^{N} \rho_j^{\beta/2-1} \tag{3.15}$$

where each ρ_j is positive and $\sum_{j=1}^{N} \rho_j = 1$.

Let us then consider the distribution of the roots of (3.13) implied by the μ_i^2 having the Dirichlet distribution implied by (3.15) [8, 1].

Proposition 3.2. *Let $\{\rho_i\}$ have the Dirichlet distribution*

$$\frac{\Gamma(N\beta/2)}{(\Gamma(\beta/2))^N} \prod_{j=1}^{N} \rho_j^{\beta/2-1}$$

and let $\{b_j\}$ be given. The roots of the random rational function

$$\sum_{j=1}^{N} \frac{\rho_j}{x - b_j},$$

denoted $\{x_1, \dots, x_{N-1}\}$ *say, have the p.d.f.*

$$\frac{\Gamma(N\beta/2)}{(\Gamma(\beta/2))^N} \frac{\prod_{1 \le j < k \le N-1} (x_j - x_k)}{\prod_{1 \le j < k \le N} (b_j - b_k)^{\beta - 1}} \prod_{j=1}^{N-1} \prod_{p=1}^{N} |x_j - b_p|^{\beta/2-1} \tag{3.16}$$

where

$$x_1 > b_1 > x_2 > b_2 > \cdots > x_{N-1} > b_N. \tag{3.17}$$

Proof. As is consistent with (3.13) write

$$\sum_{j=1}^{N} \frac{\rho_j}{x - b_j} = \frac{\prod_{l=1}^{N-1} (x - x_l)}{\prod_{l=1}^{N} (x - b_l)}.$$

For a particular j, taking the limit $x \to b_j$ shows

$$\rho_j = \frac{\prod_{l=1}^{N-1} (b_j - x_l)}{\prod_{l=1, l \ne j}^{N} (b_j - b_l)}. \tag{3.18}$$

Our task is to change variables from $\{\rho_j\}_{j=1,\dots,N-1}$ to $\{x_j\}_{j=1,\dots,N-1}$. Analogous to (3.8) we have, up to a sign

$$\bigwedge_{j=1}^{N-1} d\rho_j = \prod_{j=1}^{N-1} \rho_j \det \left[\frac{1}{b_j - x_k} \right]_{j,k=1,\dots,N-1} \bigwedge_{j=1}^{N-1} dx_j$$

and thus the corresponding Jacobian is equal to

$$\prod_{j=1}^{N-1} \rho_j \left| \frac{\prod_{1 \le j < k \le N-1} (b_k - b_j)(x_k - x_j)}{\prod_{j,k=1}^{N-1} (b_j - x_k)} \right|. \tag{3.19}$$

The result now follows immediately upon multiplying (3.19) with (3.9), and substituting for ρ_j using (3.18). □

Because, with respect to $\{x_j\}$, (3.16) is a p.d.f., integrating over the region (3.17) (R'_{N-1} say) must give unity, and so we have the integration formula

$$\int_{R'_{N-1}} dx_1 \cdots dx_{N-1} \prod_{1 \le j < k \le N-1} (x_j - x_k) \prod_{j=1}^{N-1} \prod_{p=1}^{N} |x_j - b_p|^{\beta/2-1}$$
$$= \frac{(\Gamma(\beta/2))^N}{\Gamma(N\beta/2)} \prod_{1 \le j < k \le N} (b_j - b_k)^{\beta-1}. \tag{3.20}$$

This allows us to verify the solution of the recurrence (3.12).

Proposition 3.3. *The solution of the recurrence (3.12) is given by*

$$p_N(x_1, \ldots, x_N) = \frac{1}{m_N(\beta)} \prod_{j=1}^{N} e^{-x_j^2/2} \prod_{1 \le j < k \le N} |x_k - x_j|^\beta \tag{3.21}$$

where

$$N! m_N(\beta) = (2\pi)^{N/2} \prod_{j=0}^{N-1} \frac{\Gamma(1 + (j+1)\beta/2)}{\Gamma(1 + \beta/2)}.$$

Proof. Substituting (3.21) in the r.h.s. of (3.12) gives

$$\frac{1}{\sqrt{2\pi}} \frac{1}{(\Gamma(\beta/2))^N} \frac{1}{m_N(\beta)} e^{-\frac{1}{2}\sum_{j=1}^{N+1} \lambda_j^2} \prod_{1 \le j < k \le N+1} (\lambda_j - \lambda_k)$$
$$\times \int_{R_N} da_1 \cdots da_N \prod_{1 \le j < k \le N} (a_j - a_k) \prod_{j=1}^{N+1} \prod_{p=1}^{N} |\lambda_j - a_p|^{\beta/2-1}.$$

The integral is precisely the $N \mapsto N+1$ case of (3.20). Substituting its value we obtain p_{N+1} as specified by (3.21). □

3.2. *Three term recurrence and tridiagonal matrices*

According to the working of the previous section, the characteristic polynomial $p_N(x) := \prod_{l=1}^{N}(x - x_l)$, where $\{x_j\}$ is distributed according to the Gaussian β-ensemble , satisfies the recurrence relation

$$\frac{p_{N-1}(x)}{p_N(x)} = \sum_{j=1}^{N} \frac{\rho_j}{x - x_j}$$

where

$$\rho_j \sim w_j^2/(w_1^2 + \cdots + w_N^2), \qquad w_j^2 \sim \Gamma[\beta/2, 1],$$

as well as the further recurrence relation

$$\frac{p_{N+1}(x)}{p_N(x)} = x - a - \sum_{j=1}^{N} \frac{w_j^2}{x - x_j}, \qquad a \in \mathrm{N}[0,1].$$

The two recurrences together give the random coefficient three term recurrence

$$p_{N+1}(x) = (x-a)p_N(x) - b_N^2 p_N(x). \tag{3.22}$$

The three term recurrence (3.22) occurs in the study of tridiagonal matrices. Thus consider a general real symmetric tridiagonal matrix

$$T_n = \begin{bmatrix} a_1 & b_1 & & & \\ b_1 & a_2 & b_2 & & \\ & b_2 & a_3 & & \\ & & & \vdots & b_{n-1} \\ & & & b_{n-1} & a_n \end{bmatrix}. \tag{3.23}$$

By forming $\lambda 1_n - T_n$ and expanding the determinant along the bottom row one sees

$$\det(\lambda 1_n - T_n) = (\lambda - a_n)\det(\lambda 1_{n-1} - T_{n-1}) - b_{n-1}^2 \det(\lambda 1_{n-2} - T_{n-2}).$$

Comparison with (3.22) shows the Gaussian β-ensemble is realized by the eigenvalue p.d.f. of random tridiagonal matrices with

$$a_j \sim \mathrm{N}[0,1] \qquad b_j^2 \sim \Gamma[j\beta/2, 1]. \tag{3.24}$$

This result was first obtained using different methods in [9]. The present derivation is a refinement of the approach in [20].

4. Laguerre β Ensemble

A recursive construction of the Hermite β ensemble was motivated by consideration of a recursive structure inherent in the GOE. Likewise, to motivate a recursive construction of the Laguerre β ensemble we first examine the case of the LOE. As noted in Section 2.6 this is realized by matrices $X_{(n)}^T X_{(n)}$ where $X_{(n)}$ is an $n \times N$ rectangular matrix with Gaussian entries $\mathrm{N}[0,1]$. Such matrices satisfy the recurrence

$$X_{(n+1)}^T X_{(n+1)} = X_{(n)}^T X_{(n)} + \vec{x}_{(1)} \vec{x}_{(1)}^{\,T}. \tag{4.1}$$

This suggests inductively defining a sequence of $N \times N$ positive definite matrices indexed by (n) according to

$$A_{(n+1)} = \mathrm{diag}\, A_{(n)} + \vec{x}_{(1)} \vec{x}_{(1)}^{\,T} \tag{4.2}$$

where $\operatorname{diag} A_{(n)}$ refers to the diagonal form of $A_{(n)}$ and $A_{(0)} = [0]_{N\times N}$. Noting that $A_{(n)}$ will have $N-n$ zero eigenvalues, it is therefore necessary to study the eigenvalues of the $N \times N$ matrix

$$Y := \operatorname{diag}(a_1, \dots, a_n, \underbrace{a_{n+1}, \dots, a_{n+1}}_{N-n}) + \vec{x}\vec{x}^T .$$

Since

$$\det(\lambda 1_N - Y) = \det(\lambda 1_N - A)\det(1_N - (\lambda 1_N - A)^{-1}\vec{x}\vec{x}^T)$$

it follows

$$\frac{\det(\lambda 1_N - Y)}{\det(\lambda 1_N - A)} = 1 - \sum_{j=1}^{n} \frac{x_j^2}{\lambda - a_j} - \frac{\sum_{j=n+1}^{N} x_j^2}{\lambda - a_{n+1}} . \tag{4.3}$$

One is thus led to ask about the density of zeros of the random rational function

$$1 - \sum_{j=1}^{n+1} \frac{w_j^2}{\lambda - a_j}, \tag{4.4}$$

where, since the sum of squares of Gaussian distributed variables are gamma distributed variables,

$$w_j^2 \sim \Gamma[s_j, 1]. \tag{4.5}$$

Proposition 4.1. *The zeros of the rational function (4.4) have p.d.f.*

$$\frac{1}{\Gamma(s_1)\cdots\Gamma(s_{n+1})} e^{-\sum_{j=1}^{n+1}(\lambda_j - a_j)} \times \prod_{1\le i<j\le n+1} \frac{(\lambda_i - \lambda_j)}{(a_i - a_j)^{s_i+s_j-1}} \prod_{i,j=1}^{n+1} |\lambda_i - a_j|^{s_j - 1} \tag{4.6}$$

where

$$\lambda_1 > a_1 > \lambda_2 > \cdots > \lambda_{n+1} > a_{n+1}.$$

This result can be proved [20] by following the general strategy used to establish Propositions 3.1 and 3.2.

The case of interest is

$$s_1 = \cdots = s_n = \beta/2, \quad s_{n+1} = (N-n)\beta/2, \quad a_{n+1} = 0. \tag{4.7}$$

Let us denote (4.6) with these parameters by

$$G(\{\lambda_j\}_{j=1,\dots,n+1}; \{a_j\}_{j=1,\dots,n}) .$$

Let the p.d.f. of $\{\lambda_j\}_{j=1,\dots,n+1}$ be denoted $p_{n+1}(\{a_j\})$. For $n < N$ the recursive construction of $\{A^{(n)}\}$ gives that

$$p_{n+1}(\{\lambda_j\}) = \int_{\lambda_1 > a_1 > \cdots > \lambda_{n+1} > 0} da_1 \cdots da_n \\ \times G(\{\lambda_j\}_{j=1,\dots,n+1}; \{a_j\}_{j=1,\dots,n}) p_n(\{a_j\}) \qquad (4.8)$$

subject to the initial condition $p_0 = 1$.

With $\beta = 1$ the LOE recursion (4.1) tells us that the recurrence (4.8) is satisfied by the eigenvalue p.d.f. for the non-zero eigenvalues of the Wishart matrices $X_{(n)}^T X_{(n)}$. This in turn is equal to the eigenvalue p.d.f. of the full rank matrices $X_{(n)} X_{(n)}^T$, which according to (2.18) is given by

$$p_n(\{\lambda_j\}) = \frac{1}{C_n} \prod_{l=1}^{n} \lambda_l^{(N-n-1)/2} e^{-\lambda_l} \prod_{1 \le j < k \le n} |\lambda_k - \lambda_j| \qquad (4.9)$$

(here the choice $V(x) = x$ in (2.18) has been introduced to account for the scale factor $\sigma = 1$ in the distribution $\Gamma[s_j, \sigma]$ used in (4.4)).

For general $\beta > 0$, we want to check that (4.8) has as its solution

$$p_n(\{\lambda_j\}) = \frac{1}{C_{n,\beta}} \prod_{l=1}^{n} \lambda_l^{(N-n+1)\beta/2-1} e^{-\lambda_l} \prod_{1 \le j < k \le n} |\lambda_k - \lambda_j|^{\beta}. \qquad (4.10)$$

Since

$$G(\{\lambda_j\}_{j=1,\dots,n+1}; \{a_j\}_{j=1,\dots,n}) \\ = \frac{1}{(\Gamma(\beta/2))^n \Gamma((N-n)\beta/2)} e^{-\sum_{j=1}^{n}(\lambda_j - a_j) - \lambda_{n+1}} \frac{\prod_{i<j}^{n+1}(\lambda_i - \lambda_j)}{\prod_{i<j}^{n}(a_i - a_j)^{\beta-1}} \\ \times \frac{\prod_{i=1}^{n+1} \lambda_i^{(N-n)\beta/2+1}}{a_i^{(N-n+1)\beta/2-1}} \prod_{i,j=1}^{n} |\lambda_i - a_j|^{\beta/2-1}$$

we see we need to evaluate

$$\int_{\lambda_1 > a_1 > \cdots > \lambda_{n+1} > 0} da_1 \cdots da_n \prod_{i<j}^{n}(a_i - a_j) \prod_{i,j=1}^{n} |\lambda_i - a_j|^{\beta/2-1}.$$

This is precisely the integral (3.20) with $N \mapsto n+1$, and so is equal to

$$\frac{(\Gamma(\beta/2))^{n+1}}{\Gamma((n+1)\beta/2)} \prod_{1 \le j < k \le n+1} (\lambda_j - \lambda_k)^{\beta-1},$$

leaving us with

$$\frac{C_{n+1,\beta}}{C_{n,\beta}} \frac{\Gamma(\beta/2)}{\Gamma((n+1)\beta/2)\Gamma((N-n)/\beta/2)} p_{n+1}(\{\lambda_j\}).$$

Thus (4.9) with

$$C_{n,\beta} = \prod_{k=0}^{n} \frac{\Gamma((k+1)\beta/2)\Gamma((N-k)/\beta/2)}{\Gamma(\beta/2)} \tag{4.11}$$

is indeed the solution of the recurrence (4.8).

Let $p_n(\lambda) = \prod_{l=1}^{n}(\lambda - x_l)$, where $\{x_l\}$ have the p.d.f. (4.10). We see from (4.3)–(4.5) and (4.7) that p_n satisfies the recurrence

$$\frac{p_{n+1}(\lambda)}{p_n(\lambda)} = 1 - \sum_{j=1}^{n} \frac{w_j^2}{\lambda - x_j} - \frac{w_{n+1}^2}{\lambda} \tag{4.12}$$

where

$$w_j^2 \sim \Gamma[\beta/2, 1] \ (j = 1, \ldots, n), \qquad w_{n+1}^2 \sim \Gamma[(N-n)\beta/2, 1].$$

In addition, as for the matrix M_N introduced in (3.1), the matrix $A_{(n)}$ in (4.2) must satisfy the first equality in (3.13), thus implying the companion recurrence

$$\frac{p_{n-1}(\lambda)}{p_n(\lambda)} = \sum_{j=1}^{n} \frac{\rho_j}{\lambda - x_j} \tag{4.13}$$

where

$$\rho_j \sim w_j^2/(w_1^2 + \cdots + w_n^2).$$

Comparing (4.12) and (4.13) gives the three term recurrence with random coefficients [20]

$$\lambda p_{n+1}(\lambda) = (\lambda - w_{n+1}^2) p_n(\lambda) - b_n \lambda p_{n-1}(\lambda) \tag{4.14}$$

where

$$w_{n+1}^2 \sim \Gamma[(N-n)\beta/2, 1], \qquad b_n \sim \Gamma[n\beta/2, 1].$$

5. Recent Developments

The whole topic of explicit constructions of β-random ensembles is recent, with the first paper on the subject appearing in 2002 [9]. In that work the motivation came from considerations in numerical linear algebra, whereby the form of a GOE matrix after the application of Householder transformations to tridiagonal form was sought. In the case of unitary matrices, the viewpoint of numerical linear algebra suggests seeking the Hessenberg form. Doing this [25] leads to a random matrix construction of the circular β-ensemble. Similarly, seeking the Hessenberg form of real orthogonal matrices from $O^+(N)$ leads to a random matrix construction of the Jacobi β-ensemble [25]. An alternative approach to the latter involves the cosine-sine block decomposition of unitary matrices [10].

Recurrence relations with random coefficients for the characteristic polynomials of the circular and Jacobi β-ensembles following from the underlying Hessenberg matrices are given in [25]. Using methods similar to those presented in Sections 3.1 and 4 for the Gaussian and Laguerre β-ensembles, different three term recurrences with random coefficients for the Jacobi β-ensemble and the circular β-ensemble have been given [20, 21].

Most recently [31, 11, 26] the continuum limit of various of the recurrences has been shown to be given by certain differential operators with random noise terms. In the case of the Gaussian β-ensemble this can be anticipated by viewing the corresponding tridiagonal matrix (3.23) as the discretization of a certain random Schrödinger operator [4]. This allows the scaled distributions of the particles to be described in terms of the eigenvalues of the corresponding random differential operator.

Acknowledgments

This work was supported by the Australian Research Council.

References

1. G. W. Anderson, A short proof of Selberg's generalized beta formula, *Forum Math.* **3** (1991) 415–417.
2. T. H. Baker and P. J. Forrester, The Calogero-Sutherland model and generalized classical polynomials, *Commun. Math. Phys.* **188** (1997) 175–216.
3. O. Bohigas, M. J. Giannoni and C. Schmit, Characterization of chaotic quantum spectra and universality of level fluctuation laws, *Phys. Rev. Lett.* **52** (1984) 1–4.
4. J. Breuer, P. J. Forrester and U. Smilansky, Random discrete Schrödinger operators from random matrix theory, arXiv:math-ph/0507036 (2005).
5. F. Calogero, Solution of the three-body problem in one dimension, *J. Math. Phys.* **10** (1969) 2191–2196.
6. B. Collins, Product of random projections, Jacobi ensembles and universality problems arising from free probability, *Prob. Theory Rel. Fields* **133** (2005) 315–344.
7. P. Desrosiers and P. J. Forrester, Hermite and Laguerre β-ensembles: Asymptotic corrections to the eigenvalue density, *Nucl. Phys. B* **743** (2006) 307–332.

8. A. L. Dixon, Generalizations of Legendre's formula $ke' - (k - e)k' = \frac{1}{2}\pi$, *Proc. London Math. Soc.* **3** (1905) 206–224.
9. I. Dumitriu and A. Edelman, Matrix models for beta ensembles, *J. Math. Phys.* **43** (2002) 5830–5847.
10. A. Edelman and B. D. Sutton, The beta-Jacobi matrix model, the CS decomposition, and generalized singular value problems, preprint (2006).
11. A. Edelman and B. D. Sutton, From random matrices to stochastic operators, math-ph/0607038 (2006).
12. R. J. Evans, Multidimensional beta and gamma integrals, *Contemp. Math.* **166** (1994) 341–357.
13. P. J. Forrester, Log-gases and random matrices, www.ms.unimelb.edu.au/~matpjf/matpjf.html.
14. P. J. Forrester, Selberg correlation integrals and the $1/r^2$ quantum many body system, *Nucl. Phys. B* **388** (1992) 671–699.
15. P. J. Forrester, Exact integral formulas and asymptotics for the correlations in the $1/r^2$ quantum many body system, *Phys. Lett. A* **179** (1993) 127–130.
16. P. J. Forrester, Exact results and universal asymptotics in the Laguerre random matrix ensemble, *J. Math. Phys.* **35** (1993) 2539–2551.
17. P. J. Forrester, Recurrence equations for the computation of correlations in the $1/r^2$ quantum many body system, *J. Stat. Phys.* **72** (1993) 39–50.
18. P. J. Forrester, Addendum to Selberg correlation integrals and the $1/r^2$ quantum many body system, *Nucl. Phys. B* **416** (1994) 377–385.
19. P. J. Forrester, Quantum conductance problems and the Jacobi ensemble, *J. Phys. A* **39** (2006) 6861–6870.
20. P. J. Forrester and E. M. Rains, Interpretations of some parameter dependent generalizations of classical matrix ensembles, *Probab. Theory Relat. Fields* **131** (2005) 1–61.
21. P. J. Forrester and E. M. Rains, Jacobians and rank 1 perturbations relating to unitary Hessenberg matrices, math.PR/0505552 (2005).
22. Y. V. Fyodorov, Negative moments of characteristic polynomials of random matrices: Ingham-Siegel integral as an alternative to Hubbard-Stratonovich transformation, *Nucl. Phys. B* **621** (2002) 643–674.
23. F. Haake, *Quantum Signatures of Chaos* (Springer, Berlin, 1992).
24. J. Kaneko, Selberg integrals and hypergeometric functions associated with Jack polynomials, *SIAM J. Math Anal.* **24** (1993) 1086–1110.
25. R. Killip and I. Nenciu, Matrix models for circular ensembles, *Int. Math. Res. Not.* **50** (2004) 2665–2701.
26. R. Killip and M. Stoiciu, Eigenvalue statistics for CMV matrices: From Poisson to clock via circular beta ensembles, *Duke Math. J.* **146** (2009) 361–399.
27. I. G. Macdonald, *Hall Polynomials and Symmetric Functions*, 2nd edn. (Oxford University Press, Oxford, 1995).
28. R. J. Muirhead, *Aspects of Multivariable Statistical Theory* (Wiley, New York, 1982).
29. I. Olkin, The 70th anniversary of the distribution of random matrices: A survey, *Linear Algebra Appl.* **354** (2002) 231–243.

30. C. E. Porter, *Statistical Theories of Spectra: Fluctuations* (Academic Press, New York, 1965).
31. J. Ramirez, B. Rider and B. Virag, Beta ensembles, stochastic Airy spectrum, and a diffusion, math.PR/0607331 (2006).
32. H. Risken, *The Fokker-Planck Equation* (Springer, Berlin, 1992).
33. S. H. Simon and A. L. Moustakas, Crosssover from conserving to Lossy transport in circular random-matrix ensembles, *Phys. Rev. Lett.* **96** (2006) 136805.
34. A. D. Stone, P. A. Mello, K. A. Muttalib and J.-L. Pichard, Random matrix theory and maximum entropy models for disordered conductors, in *Mesoscopic Phenomena in Solids*, eds. P. A. Lee, B. L. Altshuler and R. A. Webb (North Holland, Amsterdam, 1991), pp. 369–448.
35. B. Sutherland, Quantum many body problem in one dimension, *J. Math. Phys.* **12** (1971) 246–250.
36. J. J. M. Verbaarschot, The spectrum of the Dirac operator near zero virtuality for $n_c = 2$ and chiral random matrix theory, *Nucl. Phys. B* **426** (1994) 559–574.
37. Z. Yan, A class of generalized hypergeometric functions in several variables, *Canad. J. Math.* **44** (1992) 1317–1338.
38. K. Zyczkowski and H.-J. Sommers, Truncations of random unitary matrices, *J. Phys. A* **33** (2000) 2045–2057.

FUTURE OF STATISTICS

Zhidong Bai* and Shurong Zheng†
*National University of Singapore
6 Science Drive 2, 117546, Singapore
†Northeast Normal University
Changchun, Jilin 130024, P. R. China
E-mails: *stabaizd@nus.edu.sg
†zhengsr@nenu.edu.cn

With the rapid development of modern computer science, large dimensional data analysis has become more and more important and has therefore received increasing attention from statisticians. In this note, we shall illustrate the difference between large dimensional data analysis and classical data analysis by means of some examples, and we show the importance of random matrix theory and its applications to large dimensional data analysis.

1. Introduction

What are the future aspects of modern statistics and in which direction will it develop? To answer this, we shall have a look at what has influenced statistical research in recent decades. We strongly believe that, in every discipline, the most impacting factor has been — and still is — the rapid development and wide application of computer technology and computing sciences. It has become possible to collect, store and analyze huge amounts of data of large dimensionality. As a result, more and more measurements are collected with large dimension, e.g. data in curves, images and movies, and statisticians have to face the task of analyzing these data. But computer technology also offers big advantages. We are now in a position to do many things that were not possible 20 years ago, such as making spectral decompositions of a matrix of order 1000×1000, searching patterns in a DNA sequences and much more. However, it also confronts us with the big challenge that classical limit theorems are no longer suitable to deal with large dimensional data and we have to develop new limit theorems to cope with this. As a result, many statisticians have now become interested in this research topic.

Typically, large dimensional problems involve a "large" dimension p and a "small" sample size n. However, in a real problem, they both are given integers. It is then natural to ask for which size the dimension p has to be taken as fixed or tending to infinity and what we should do if we cannot justify "p is fixed". Is it reasonable to claim "p is fixed" if the ratio of dimension and sample size p/n is small, say, less than 0.001? If we cannot say "p is fixed", can any limit theorems be used for large dimensional data analysis?

To discuss these questions, we shall provide some examples of multivariate analysis. We illustrate the difference between traditional tests and the new approaches of large dimensional data by considering tests on the difference of two population means and tests on the equality of a population covariance matrix and a given matrix. By means of simulations, we will show how the new approaches are superior to the traditional ones.

At present, large dimensional random matrix theory (RMT) is the only systematic theory which is applicable to large dimensional problems. The RMT is different from the classical limit theories because it is built on the assumption that $p/n \to y > 0$ regardless what y is, provided where it is applicable, say $y \in (0,1)$ for T^2 statistic. The RMT shows that the classical limit theories behave very poorly or are even inapplicable to large-dimensional problems, especially when the dimension is growing proportionally with the sample size [see Dempster (1958), Bai (1993a,b, 1999), Bai and Saranadasa (1996), Bai and Silverstein (2004, 2006), Bai and Yin (1993), Bai, Yin and Krishnaiah (1988)]. In this paper, we will show how to deal with large dimensional problems with the help of RMT, especially the CLT of Bai and Silverstein (2004).

2. A Multivariate Two-Sample Problem

In this section, we revisit the T^2 test for the two-sample problem. Suppose that $\mathbf{x}_{i,j} \sim N_p(\boldsymbol{\mu}_i, \Sigma)$, $j = 1, \ldots, N_i$, $i = 1, 2$, are two independent samples. To test the hypotheses $H_0 : \boldsymbol{\mu}_1 = \boldsymbol{\mu}_2$ vs $H_1 : \boldsymbol{\mu}_1 \neq \boldsymbol{\mu}_2$, traditionally one uses Hotelling's famous T^2-test which is defined by

$$T^2 = \eta(\bar{\mathbf{x}}_1 - \bar{\mathbf{x}}_2)' A^{-1} (\bar{\mathbf{x}}_1 - \bar{\mathbf{x}}_2), \tag{2.1}$$

where $\bar{\mathbf{x}}_i = \frac{1}{N_i}\sum_{j=1}^{N_i}\mathbf{x}_{i,j}$, $i = 1, 2$, $A = \sum_{i=1}^{2}\sum_{j=1}^{N_i}(\mathbf{x}_{i,j} - \bar{\mathbf{x}}_i)(\mathbf{x}_{i,j} - \bar{\mathbf{x}}_i)'$ and $\eta = n\frac{N_1N_2}{N_1+N_2}$ with $n = N_1 + N_2 - 2$. It is well known that, under the null hypothesis, the T^2 statistic has an F distribution with degrees of freedom p and $n - p + 1$.

The advantages of the T^2-test include the properties that it is invariant under affine transformations, has an exact known null distribution, and is most powerful when the dimension of data is sufficiently small compared to its sample size. However, Hotelling's test has the serious defect that the T^2 statistic is undefined when the dimension of data is greater than the sample degrees of freedom. Looking for remedies, Chung and Fraser (1958) proposed a nonparametric test and Dempster (1958, 1960) discussed the so-called "non-exact" significance test (NET). Dempster (1960) also considered the so-called randomization test. Not only being a remedy when the T^2 is undefined, Bai and Saranadasa (1996) also found that, even if T^2 is well defined, the NET is more powerful than the T^2 test when the dimension is "close to" the sample degrees of freedom. Both, the T^2 test and Dempster's NET, strongly rely on the normality assumption. Moreover, Dempster's non-exact test statistic involves a complicated estimation of r, the degrees of freedom for the chi-square approximation. To simplify the testing procedure, a new method, the Asymptotic Normality Test (ANT), is proposed in Bai and Sarahadasa (1996). It is proven there that the asymptotic power of ANT is equivalent to that of Dempster's NET. Simulation results further show that the new approach is slightly more powerful than Dempster's NET. We believe that the estimation of r and its rounding to an integer in Dempster's procedure may cause an error of order $O(1/n)$. This might indicate that the new approach is superior to Dempster's test in the second order term in some Edgeworth-type expansions (see Babu and Bai (1993) and Bai and Rao (1991) for reference of Edgeworth expansions).

2.1. *Asymptotic power of T^2 test*

The purpose of this section is to investigate the asymptotic power of Hotelling's test when $p/n \to y \in (0, 1)$ and to compare it with other NETs given in later sections. To derive the asymptotic power of Hotelling's test, we first derive an asymptotic expression for the threshold of the test. It is well known that under the null hypothesis, $\frac{n-p+1}{np}T^2$ has an F-distribution with degrees of freedom p and $n - p + 1$. Let the significance level be chosen as α and the threshold be denoted by $F_\alpha(p, n - p + 1)$. By elementary calculations, we can prove the following.

Lemma 1. *We have* $\frac{p}{n-p+1}F_\alpha(p, n-p+1) = \frac{y_n}{1-y_n} + \sqrt{2y/(1-y)^3 n} z_\alpha + o(1/\sqrt{n})$*, where* $y_n = p/n$, $\lim_{n\to\infty} y_n = y \in (0,1)$ *and* z_α *is the* $1-\alpha$ *quantile of the standard normal distribution.*

Now, to describe the asymptotic behavior of the T^2 statistic under the alternative hypothesis H_1, one can easily show that the distribution of the T^2 statistic is the same as of

$$(\mathbf{w} + \tau^{-1/2}\boldsymbol{\delta})' U^{-1} (\mathbf{w} + \tau^{-1/2}\boldsymbol{\delta}), \tag{2.2}$$

where $\boldsymbol{\delta} = \Sigma^{-1/2}(\boldsymbol{\mu}_1 - \boldsymbol{\mu}_2)$, $U = \sum_{i=1}^n \mathbf{u}_i \mathbf{u}_i'$, $\mathbf{w} = (w_1, \ldots, w_p)'$ and $\mathbf{u}_i, i = 1, \ldots, n$ are i.i.d. $N(0, I_p)$ random vectors and $\tau = \frac{N_1+N_2}{N_1 N_2}$, and Σ is covariance matrix of the riginal population. Denote the spectral decomposition of U^{-1} by $O\text{diag}[d_1, \ldots, d_p]O'$ with eigenvalues $d_1 \geq \cdots \geq d_p > 0$. Then, (2.2) becomes

$$(O\mathbf{w} + \tau^{-1/2}\|\boldsymbol{\delta}\|\mathbf{v})' \text{diag}[d_1, \ldots, d_p](O\mathbf{w} + \tau^{-1/2}\|\boldsymbol{\delta}\|\mathbf{v}), \tag{2.3}$$

where $\mathbf{v} = O\boldsymbol{\delta}/\|\boldsymbol{\delta}\|$. Since U has the Wishart distribution $W(n, I_p)$, the orthogonal matrix O has the Haar distribution on the group of all orthogonal p-matrices, and hence the vector $\mathbf{v}$ is uniformly distributed on the unit p-sphere. Note that the conditional distribution of $O\mathbf{w}$ given O is $N(0, I_p)$, the same as that of $\mathbf{w}$, which is independent of O. This shows that $O\mathbf{w}$ is independent of $\mathbf{v}$. Therefore, replacing $O\mathbf{w}$ in (2.3) by $\mathbf{w}$ does not change the joint distribution of $O\mathbf{w}$, $\mathbf{v}$ and the d_i's. Consequently, T^2 has the same distribution as

$$\Omega_n = \sum_{i=1}^p (w_i^2 + 2w_i v_i \tau^{-1/2}\|\boldsymbol{\delta}\| + \tau^{-1}\|\boldsymbol{\delta}\|^2 v_i^2) d_i, \tag{2.4}$$

where $\mathbf{v} = (v_1, \ldots, v_p)'$ is uniformly distributed on the unit sphere of R^p and is independent of $\mathbf{w}$ and the d_i's.

Lemma 2. *Using the above notation, we have* $\sqrt{n}\left(\sum_{i=1}^p d_i - \frac{y_n}{1-y_n}\right) \to 0$, *and* $n\sum_{i=1}^p d_i^2 \to \frac{y}{(1-y)3}$ *in probability.*

Now we are in a position to express the approximation of the power function of Hotelling's test.

Theorem 3. *If* $y_n = p/n \to y \in (0,1)$, $N_1/(N_1+N_2) \to \kappa \in (0,1)$ *and* $\|\boldsymbol{\delta}\| = o(1)$, *then*

$$\beta_H(\boldsymbol{\delta}) - \Phi\left(-z_\alpha + \sqrt{\frac{n(1-y)}{2y}}\kappa(1-\kappa)\|\boldsymbol{\delta}\|^2\right) \to 0, \tag{2.5}$$

where $\beta_H(\boldsymbol{\delta})$ is the power function of Hotelling's test and Φ is the distribution function of standard normal random variable.

Remark 4. If the alternative hypothesis is considered in limit theorems, it is typically assumed that $\sqrt{n}\|\boldsymbol{\delta}\|^2 \to a > 0$. Under this additional assumption, it follows from (2.5) that the limiting power of Hotelling's test is given by $\beta_H(\boldsymbol{\delta}) - \Phi(-z_\alpha + \sqrt{\frac{(1-y)}{2y}}\kappa(1-\kappa)a)$. This formula shows that the limiting power of Hotelling's test is slowly increasing for y close to 1 as the non-central parameter a increases.

2.2. *Dempster's NET*

Dempster (1958, 1960) proposed a non-exact test for the hypothesis H_0 with the dimension of data possibly greater than the sample degrees of freedom. Let us briefly describe his test. Denote $N = N_1 + N_2$, $\mathbf{X}' = (\mathbf{x}_{11}, \mathbf{x}_{12}, \ldots, \mathbf{x}_{1N_1}; \mathbf{x}_{21}, \ldots, \mathbf{x}_{2N_2})$ and by $H' = (\frac{1}{\sqrt{N}}\mathbf{J}_N, (\sqrt{\frac{N_2}{N_1 N}}\mathbf{J}'_{N_1}, -\sqrt{\frac{N_1}{N_2 N}}\mathbf{J}'_{N_2})', \mathbf{h}_3, \ldots, \mathbf{h}_N)$ a suitably chosen orthogonal matrix, where $\mathbf{J}_d$ is a d-dimensional column vector of 1's. Let $\mathbf{Y} = H\mathbf{X} = (\mathbf{y}_1, \ldots, \mathbf{y}_N)'$. Then, the vectors $\mathbf{y}_1, \ldots, \mathbf{y}_N$ are independent normal random vectors with $E(\mathbf{y}_1) = (N_1\boldsymbol{\mu}_1 + N_2\boldsymbol{\mu}_2)/\sqrt{N}$, $E(\mathbf{y}_2) = \tau^{-1/2}(\boldsymbol{\mu}_1 - \boldsymbol{\mu}_2)$, $E(\mathbf{y}_j) = 0$, for $3 \le j \le N$, $\mathrm{Cov}(\mathbf{y}_j) = \Sigma$, $1 \le j \le N$. Dempster proposed the NET statistic $F = Q_2/(\sum_{i=3}^N Q_i)/n$, where $Q_i = \mathbf{y}'_i\mathbf{y}_i$, $n = N - 2$. He used the so-called χ^2-approximation technique, assuming Q_i is approximately distributed as $m\chi^2_r$, where the parameters m and r may be found by the method of moments. Then, the distribution of F is approximately $F_{r,nr}$. But generally the parameter r (its explicit form is given in (2.8) below) is unknown. Dempster estimated r by either of the following two ways.

Approach 1: $\hat{r}$ is the solution of the equation

$$t = \left(\frac{1}{\hat{r}_1} + \frac{1 + \frac{1}{n}}{3\hat{r}_1^2}\right)(n-1). \tag{2.6}$$

Approach 2: $\hat{r}$ is the solution of the equation

$$t + w = \left(\frac{1}{\hat{r}_2} + \frac{1 + \frac{1}{n}}{3\hat{r}_2^2}\right)(n-1) + \left(\frac{1}{\hat{r}_2} + \frac{3}{2\hat{r}_2^2}\right)\binom{n}{2}, \tag{2.7}$$

where $t = n[\ln(\frac{1}{n}\sum_{i=3}^{N} Q_i)] - \sum_{i=3}^{N} \ln Q_i$, $w = -\sum_{3\le i<j\le N} \ln\sin^2\theta_{ij}$ and θ_{ij} is the angle between the vectors of $\mathbf{y}_i$, $\mathbf{y}_j$, $3 \le i < j \le N$. Dempster's test is then to reject H_0 if $F > F_\alpha(\hat{r}, n\hat{r})$.

By elementary calculus, we have

$$r = \frac{(\mathrm{tr}(\Sigma))^2}{\mathrm{tr}(\Sigma^2)} \quad \text{and} \quad m = \frac{\mathrm{tr}(\Sigma^2)}{\mathrm{tr}\,\Sigma}. \tag{2.8}$$

From (2.8) and the Cauchy-Schwarz inequality, it follows that $r \le p$. On the other hand, under regular conditions, both $\mathrm{tr}(\Sigma)$ and $\mathrm{tr}(\Sigma^2)$ are of the order $O(n)$, and hence, r is of the same order. Under wider conditions (2.12) and (2.13) given in Theorem 6 below, it can be proven that $r \to \infty$. Further, we may prove that $t \sim (n/r)N(1, \frac{1}{\sqrt{n}})$ and $w \sim \frac{n(n-1)}{2r}N(1, \frac{4}{n(n-1)} + \frac{8}{nr})$. From these estimates, one may conclude that both $\hat{r}_1$ and $\hat{r}_2$ are ratio-consistent (in the sense that $\hat{r}/r \to 1$). Therefore, the solutions of equations (2.6) and (2.7) should satisfy

$$\hat{r}_1 = \frac{n}{t} + O(1) \tag{2.9}$$

and

$$\hat{r}_2 = \frac{1}{w}\binom{n}{2} + O(1), \tag{2.10}$$

respectively. Since the random effect may cause an error of order $O(1)$, one may simply choose the estimates of r as $\frac{n}{t}$ or $\frac{1}{w}\binom{n}{2}$.

To describe the asymptotic power function of Dempster's NET, we assume that $p/n \to y > 0$, $N_1/N \to \kappa \in (0,1)$ and that the parameter r is known. The reader should note that the limiting ratio y is allowed to be greater than one in this case. When r is unknown, substituting r by the estimators $\hat{r}_1$ or $\hat{r}_2$ may cause an error of high order smallness in the approximation of the power function of Dempster's NET. Similar to Lemma 1 one may show the following.

Lemma 5. *When $n, r \to \infty$,*

$$F_\alpha(r, nr) = 1 + \sqrt{2/r}\,z_\alpha + o(1/\sqrt{r}). \tag{2.11}$$

Then we have the following approximation of the power function of Dempster's NET.

Theorem 6. *If*

$$\boldsymbol{\mu}'\Sigma\boldsymbol{\mu} = o(\tau\,\mathrm{tr}\,\Sigma^2), \tag{2.12}$$

$$\lambda_{\max} = o(\sqrt{\mathrm{tr}\,\Sigma^2}), \tag{2.13}$$

and r is known, then

$$\beta_D(\boldsymbol{\mu}) - \Phi\Big(- z_\alpha + \frac{n\kappa(1-\kappa)\|\boldsymbol{\mu}\|^2}{\sqrt{\operatorname{tr}\Sigma^2}}\Big) \to 0, \tag{2.14}$$

where $\boldsymbol{\mu} = \boldsymbol{\mu}_1 - \boldsymbol{\mu}_2$.

Remark 7. In usual cases when considering the asymptotic power of Dempster's test, the quantity $\|\boldsymbol{\mu}\|^2$ is typically assumed to have the same order as $1/\sqrt{n}$ and $\operatorname{tr}(\Sigma^2)$ to have the order n. Thus, the quantities $n\|\boldsymbol{\mu}\|^2/\sqrt{\operatorname{tr}\Sigma^2}$ and $\sqrt{n}\|\boldsymbol{\delta}\|^2$ are both bounded away from zero and infinity. The expression of the asymptotic power of Hotelling's test involves a factor $\sqrt{1-y}$ which disappears in the expression of the asymptotic power of Dempster's test. This reveals the reason why the power of the Hotelling test increases much slower than that of the Dempster test as the non-central parameter increases if y is close to one.

2.3. *Bai and Saranadasa's ANT*

In this section, we describe the results for Bai and Saranadasa's ANT. We shall not assume the normality of the underlying distributions. We assume:

(a) $\mathbf{x}_{ij} = \Gamma\mathbf{z}_{ij} + \boldsymbol{\mu}_i; j = 1, \ldots, N_i, i = 1, 2$, where Γ is a $p \times m$ matrix ($m \leq \infty$) with $\Gamma\Gamma' = \Sigma$ and $\mathbf{z}_{ij}$ are i.i.d. random m-vectors with independent components satisfying $E\mathbf{z}_{ij} = 0$, $\operatorname{Var}(\mathbf{z}_{ij}) = I_m$, $E\mathbf{z}_{ijk}^4 = 3 + \Delta < \infty$ and $E\prod_{k=1}^m z_{ijk}^{\nu_k} = 0$ (and 1) when there is at least one $\nu_k = 1$ (there are two ν_k's equal to 2, correspondingly), whenever $\nu_1 + \cdots + \nu_m = 4$.
(b) $p/n \to y > 0$ and $N_1/N \to \kappa \in (0, 1)$.
(c) (2.12) and (2.13) are true.

Here and later, it should be noted that all random variables and parameters depend on n. For simplicity we omit the subscript n from all random variables except those statistics defined later.

Now, we begin to construct the ANT proposed in Bai and Saranadasa (1996). Consider the statistic

$$M_n = (\bar{\mathbf{x}}_1 - \bar{\mathbf{x}}_2)'(\bar{\mathbf{x}}_1 - \bar{\mathbf{x}}_2) - \tau \operatorname{tr} S_n, \tag{2.15}$$

where $S_n = \frac{1}{n}A$, $\bar{\mathbf{x}}_1$, $\bar{\mathbf{x}}_2$ and A are defined in previous sections. Under H_0, we have $EM_n = 0$. If Conditions (a)–(c) hold, it can be proven that, under H_0,

$$Z_n = \frac{M_n}{\sqrt{\operatorname{Var} M_n}} \to N(0, 1), \quad \text{as } n \to \infty. \tag{2.16}$$

If the underlying distributions are normal, then, again under H_0, we have

$$\sigma_M^2 := \operatorname{Var} M_n = 2r^2\left(1+\frac{1}{n}\right)\operatorname{tr}\Sigma^2. \tag{2.17}$$

If the underlying distributions are not normal but satisfy Conditions (a)–(c), one may show that

$$\operatorname{Var} M_n = \sigma_M^2(1+o(1)). \tag{2.18}$$

Hence (2.16) is still true if the denominator of Z_n is replaced by σ_M. Therefore, to complete the construction of the ANT statistic, we only need to find a ratio-consistent estimator of $\operatorname{tr}(\Sigma^2)$ and substitute it into the denominator of Z_n. It seems that a natural estimator of $\operatorname{tr}\Sigma^2$ should be $\operatorname{tr} S_n^2$. However, unlike the case where p is fixed, $\operatorname{tr} S_n^2$ is generally neither unbiased nor ratio-consistent even under the normality assumption. If $nS_n \sim W_p(n,\Sigma)$, it is routine to verify that

$$B_n^2 = \frac{n^2}{(n+2)(n-1)}\left(\operatorname{tr} S_n^2 - \frac{1}{n}(\operatorname{tr} S_n)^2\right)$$

is an unbiased and ratio-consistent estimator of $\operatorname{tr}\Sigma^2$. Here, it should be noted that $\operatorname{tr} S_n^2 - \frac{1}{n}(\operatorname{tr} S_n)^2 \geq 0$, by the Cauchy-Schwarz inequality. It is not difficult to prove that B_n^2 is also a ratio-consistent estimator of $\operatorname{tr}\Sigma^2$ under Conditions(a)–(c). Replacing $\operatorname{tr}\Sigma^2$ in (2.17) by the ratio-consistent estimator B_n^2, we obtain the ANT statistic

$$\begin{aligned} Z &= \frac{(\bar{\mathbf{x}}_1-\bar{\mathbf{x}}_2)'(\bar{\mathbf{x}}_1-\bar{\mathbf{x}}_2) - \tau \operatorname{tr} S_n}{\tau\sqrt{\frac{2(n+1)n}{(n+2)(n-1)}\left(\operatorname{tr} S_n^2 - n^{-1}(\operatorname{tr} S_n)^2\right)}} \\ &= \frac{\frac{N_1N_2}{N}(\bar{\mathbf{x}}_1-\bar{\mathbf{x}}_2)'(\bar{\mathbf{x}}_1-\bar{\mathbf{x}}_2) - \operatorname{tr} S_n}{\sqrt{\frac{2(n+1)}{n}}B_n} \to N(0,1). \end{aligned} \tag{2.19}$$

Due to (2.19) the test rejects H_0 if $Z > z_\alpha$. Regarding the asymptotic power of our new test, we have the following theorem.

Theorem 8. *Under Conditions (a)–(c),*

$$\beta_{BS}(\boldsymbol{\mu}) - \Phi\left(-z_\alpha + \frac{n\kappa(1-\kappa)\|\boldsymbol{\mu}\|^2}{\sqrt{2\operatorname{tr}\Sigma^2}}\right) \to 0. \tag{2.20}$$

2.4. *Conclusions and simulations*

Comparing Theorems 3, 6 and 8, we find that, from the point of view of large sample theory, Hotelling's test is less powerful than the other two tests when y is close to one and that the latter two tests have the same asymptotic power function. Our simulation results show that even for moderate sample and dimension sizes, Hotelling's test is still less powerful than the other two tests when the underlying covariance structure is reasonably regular (i.e. the structure of Σ does not cause too large a difference between $\boldsymbol{\mu}'\Sigma^{-1}\boldsymbol{\mu}$ and $\sqrt{n}\|\boldsymbol{\mu}\|^2/\sqrt{\operatorname{tr}(\Sigma^2)}$), whereas the Type I error does not change much in the latter two tests. It would not be hard to see that, using the approach of this paper, one may easily derive similar results for the one-sample problem, namely, Hotelling's test is less powerful than NET and ANT, when the dimension of data is large. Now, let us shed some light on this phenomenon. The reason Hotelling's test being less powerful is the "inaccuracy" of the estimator of the covariance matrix. Let $\mathbf{x}_1, \ldots, \mathbf{x}_n$ be i.i.d. random p-vectors of mean 0 and variance-covariance matrix I_p. By the law of large numbers, the sample covariance matrix $S_n = n^{-1}\sum_{i=1}^n \mathbf{x}_i\mathbf{x}_i'$ should be "close" to the identity I_p with an error of the order $O_p(1/\sqrt{n})$ when p is fixed. However, when p is proportional to n (say $p/n \to y \in (0,1)$), the ratio of the largest and the smallest eigenvalues of S_n tends to $(1+\sqrt{y})^2/(1-\sqrt{y})^2$ (see e.g. Bai, Silverstein and Yin (1988), Bai and Yin (1993), Geman (1980), Silverstein (1985) and Yin, Bai and Krishnaiah (1988)). More precisely, in the theory of spectral analysis of large dimensional random matrices, it has been proven that the empirical distribution of the eigenvalues of S_n tends to a limiting distribution spreading over $[(1-\sqrt{y})^2, (1+\sqrt{y})^2]$ as $n \to \infty$. (see e.g. Jonsson (1982), Wachter (1978), Yin (1986) and Yin, Bai and Krishnaiah (1983)). This implies that S_n is not close to I_p. Especially when y is "close" to one, then S_n has many small eigenvalues and hence S_n^{-1} has many huge eigenvalues. This will cause the deficiency of the T^2 test. We believe that in many other multivariate statistical inferences with an inverse of a sample covariance matrix involved the same phenomenon should exist (as another example, see Saranadasa (1993)). Let us now explain our quotation-marked "close" to one. Note that the limiting ratio between the largest and smallest eigenvalues of S_n tends to $(1+\sqrt{y})^2/(1-\sqrt{y})^2$. For our simulation example, $y = 0.93$ and the ratio of the extreme eigenvalues is about 3039. This is very serious. Even for y as small as 0.1 or 0.01, the ratio can be as large as 3.705 and 1.494, which shows that it is not even necessary to require the dimension of data to be very close to the degrees of freedom to make the effect of high dimension visible. In fact, this has been shown by our simulation for $p = 4$.

Dempster's test statistic depends on the choice of vectors $\mathbf{h}_3, \mathbf{h}_4, \ldots, \mathbf{h}_N$ because different choices of these vectors would result in different estimates of the parameter r. On the other hand, the estimation of r and the rounding of the estimates may cause an error (probably an error of second order smallness) in Dempster's test. Thus, we conjecture that our new test can be more powerful than Dempster's in their second terms of an Edgeworth type expansion of their power functions. This conjecture was strongly supported by our simulation results. Because our test statistic is mathematically simple, it is not difficult to get an Edgeworth expansion by using the results obtain in Babu and Bai (1993), Bai and Rao (1991) or Bhattacharya and Ghosh (1978). It seems difficult to get a similar expansion for Dempster's test due to his complicated estimation of r.

We conducted a simulation study to compare the powers of the three tests for both normal and non-normal cases with the dimensions $N_1 = 25$, $N_2 = 20$, and $p = 40$. For the non-normal case, observations were generated by the following moving average model. Let $\{U_{ijk}\}$ be a set of independent gamma variables with shape parameter 4 and scale parameter 1. Define

$$x_{ijk} = U_{ijk} + \rho U_{i,j+1,k} + \varepsilon_{jk}; (k = 1, \ldots, p, j = 1, \ldots, N_i, i = 1, 2),$$

where ρ and the μ's are constants. Under this model, $\Sigma = (\sigma_{ij})$ with $\sigma_{ii} = 4(1 + \rho^2)$, $\sigma_{i,i+1} = 4$ and $\sigma_{ij} = 0$ for $|i - j| > 1$. For the normal case, the covariance matrices were chosen to be $\Sigma = I_p$ and $\Sigma = (1 - \rho)I_p + \rho J_p$, with $\rho = 0.5$, where J is a $p \times p$ matrix with all entries 1. A simulation was also conducted for small p (chosen as $p = 4$). The tests were made for size $\alpha = 0.05$ with 1000 repetitions. The power was evaluated at standard parameter $\eta = \|\boldsymbol{\mu}_1 - \boldsymbol{\mu}_2\|^2/\sqrt{\operatorname{tr}\Sigma^2}$. The simulation for the non-normality case was conducted for $\rho = 0$, 0.3, 0.6 and 0.9 (Figure 1). All three tests have almost the same significance level. Under the alternative hypothesis, the power curves of Dempster's test and our test are rather close but that of our test is always higher than Dempster's test. Theoretically, the power function for Hotelling's test should increase very slowly when the noncentral parameter increases. This is also demonstrated by our simulation results. The reader should note that there are only 1000 repetitions for each value of the noncentral parameter in our simulation which may cause an error of $1/\sqrt{1000} = 0.0316$ by the Central Limit Theorem. Hence, it is not surprising that the simulated power function of the Hotelling's test, whose magnitude is only around 0.05, seems not to be increasing at some points

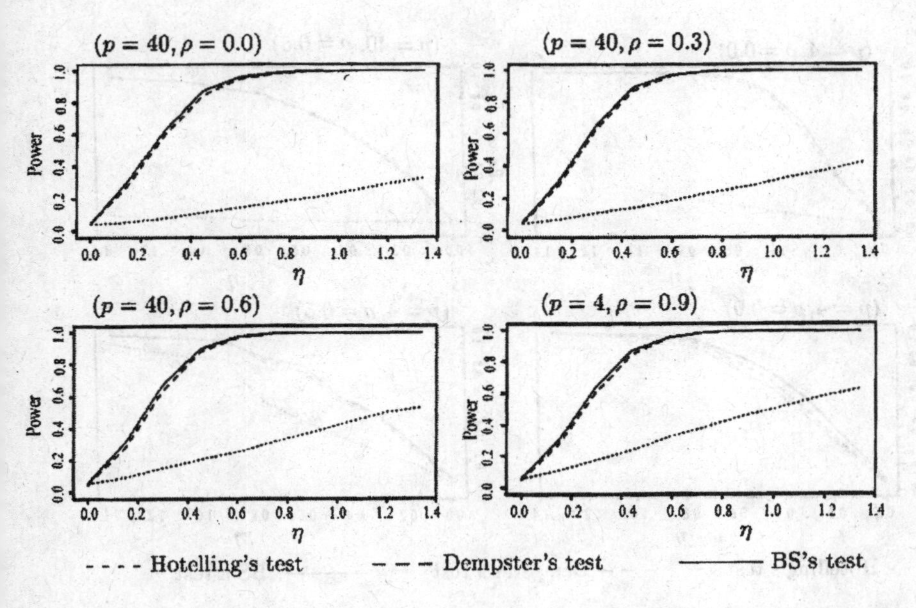

Fig. 1. Simulated powers of the three tests with multivariate Gamma distributions.

of the noncentral parameter. Similar tables are presented for the normal case (Figure 2). For higher dimension cases the power functions of Dempster's test and our test are almost the same, and our method is not worse than Hotelling's test even for $p = 4$.

3. A Likelihood Ratio Test on Covariance Matrix

In multivariate analysis, the second important test is about the covariance matrix. Assume that $\mathbf{x}_i = (x_{1i}, \ldots, x_{pi})'$ is a sample from a multivariate normal population with mean vector $\mathbf{0}_p$ and variance-covariance matrix $\Sigma_{p\times p}$ for $i = 1, \ldots, n$. Now, consider the hypotheses

$$H_0 : \Sigma = I_{p\times p} \quad v.s. \quad H_1 : \Sigma \neq I_{p\times p}. \tag{3.1}$$

3.1. *Classical tests*

It is known that the sufficient and complete statistic for Σ is the sample covariance matrix which is defined by

$$S_n = \frac{1}{n}\sum_{i=1}^{n} \mathbf{x}_i\mathbf{x}_i'. \tag{3.2}$$

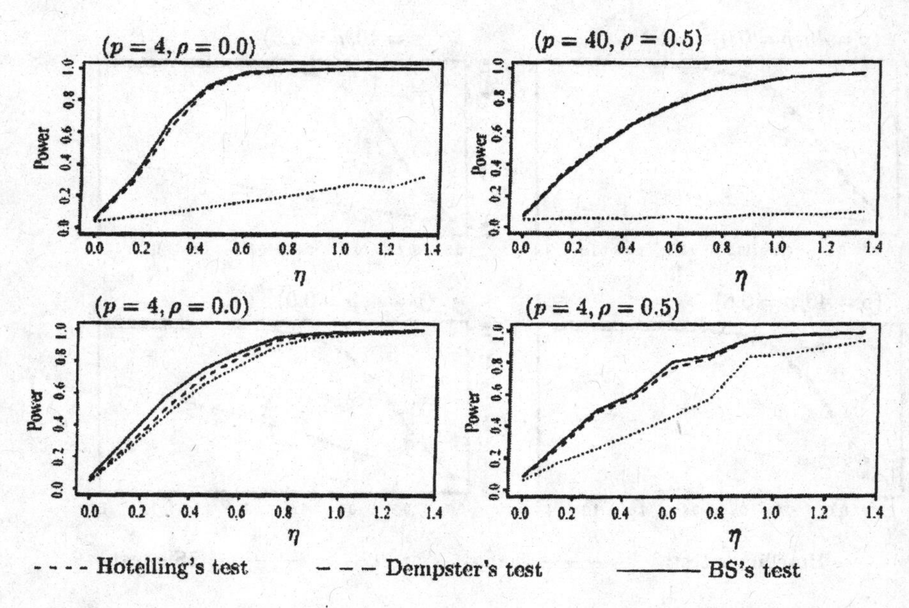

Fig. 2. Simulated powers of the three tests with multivariate normal distributions.

A classical test statistic is the log-determinant of S_n,

$$T_1 = \log(|S_n|). \tag{3.3}$$

Another important test is the likelihood ratio test (LRT) whose test statistic is given by

$$T_2 = n(\mathrm{tr}(S_n) - \log(|S_n|) - p) \tag{3.4}$$

(see Anderson 1984). To test H_0, we have the following limiting theorem.

Theorem 9. *Under null hypothesis, for any fixed p, as $n \to \infty$,*

$$\sqrt{\frac{n}{2p}}T_1 \xrightarrow{D} N(0,1),$$

$$T_2 \xrightarrow{D} \chi^2_{p(p+1)/2}.$$

The limiting distributions of T_1 and T_2 in Theorem 9 are valid even without the normality assumption under existence of the 4th moment. It only needs the assumption of fixed dimension, or even weaker, $p/n \to 0$. However, both p and n are given numbers in a real testing problem. How could one justify the assumption $p/n \to 0$? That is, for what pairs of (p, n) can Theorem 9 be applied to the tests?

For large-dimensional problems, the approximation of T_1 and T_2 by their respective asymptotic limits can cause severe errors. Let us have a look at the following simulation results given in the following table in which empirical Type I errors are listed with significance level $\alpha = 5\%$ and 40 000 simulations.

Type I errors

$n = 500$					
p	5	10	50	100	300
T_1	0.0567	0.0903	1.0	1.0	1.0
T_2	0.0253	0.0273	0.1434	0.9523	1.0
$n = 1000$					
p	5	10	50	100	300
T_1	0.0530	0.0666	0.9830	1.0	1.0
T_2	0.0255	0.0266	0.0678	0.4221	1.0
$p/n = 0.05$					
(n, p)	(250, 12)	(500, 25)	(1000, 50)	(2000, 100)	(6000,300)
T_1	0.1835	0.5511	0.9830	1.0	1.0
T_2	0.0306	0.0417	0.0678	0.1366	0.7186

The simulation results show that Type I errors for the classical methods T_1 and T_2 are close to 1 as $p/n \to y \in (0, 1)$ or p/n is large. It shows that the classical methods T_1 and T_2 behave very poorly and are even inapplicable for the testing problems with large dimension or dimension increasing with sample size.

Bai and Silverstein (2004) have revealed the reason of the above phenomenon. They show that, with probability 1,

$$T_1 = \sqrt{\frac{n}{2p}} \cdot \log(|S_n|) \to -\infty \quad \text{as} \quad p/n \to y \in (0, 1). \tag{3.5}$$

We can similarly show that

$$T_2 = n \cdot (\operatorname{tr}(S_n) - \log(|S_n|) - p) \to -\infty \quad \text{as} \quad p/n \to y \in (0, 1). \tag{3.6}$$

These two results show that Theorem 9 is not applicable when p is large and we have to seek for new limit theorems to test the Hypothesis 3.1.

3.2. *Random matrix theory*

In this section, we shall cite some important results of RMT related to our work. For more details on RMT, the reader is referred to Bai (1999), Bai and Silverstein (1998, 2004), Bai, Yin and Krishnaiah (1983), Marčenko and Pastur (1967).

Before dealing with the large-dimensional testing problem (3.1), we first introduce some basic concepts, notations and present some well-known results. To begin with, suppose that $\{x_{ij},\ i,j = 1,2,\dots\}$ is a double array of i.i.d. complex random variables with mean zero and variance 1. Write $\mathbf{x}_k = (x_{1k},\dots,x_{pk})'$ for $k = 1,\dots,n$ and

$$\mathbf{B}_p = T_p^{1/2}\left(\frac{1}{n}\sum_{k=1}^{n}\mathbf{x}_k\mathbf{x}_k^*\right)T_p^{1/2},$$

where $Ex_{11} = 0$, $E|x_{11}|^2 = 1$, $T_p^{1/2}$ is $p\times p$ random non-negative definite Hermitian with $(\mathbf{x}_1,\dots,\mathbf{x}_n)$ and $T_p^{1/2}$ being independent. Let F^A denote the *empirical spectral distribution* (ESD) of the eigenvalues of the square matrix A, that is, if A is $p\times p$, then

$$F^A(x) = \frac{(\text{number of eigenvalues of } A \le x)}{p}.$$

Silverstein (1995) proved that under certain conditions, with probability 1, $F^{\mathbf{B}_p}$ tends to a limiting distribution, called the limiting spectral distribution (LSD). To describe his result, we define the Stieltjes transform for the c.d.f. G by

$$m_G(z) \equiv \int \frac{1}{\lambda - z}dG(\lambda), \qquad z\in\mathbb{C}^+ = \{z:\ z\in\mathbb{C}, \Im(z) > 0\}. \tag{3.7}$$

Let $H_p = F^{T_p}$ and H denote the ESD and *limiting spectral distribution* (LSD) of T_p, respectively. Also, let $F^{\{y,H\}}$ denote the LSD of F^{B_p}. Further, let $F^{\{y_p,H_p\}}$ denote the LSD $F^{\{y,H\}}$ with $y = y_p$ and $H = H_p$.

Let $\underline{m}(\cdot)$ and $m_{\underline{F}^{\{y_p,H_p\}}}(\cdot)$ denote the Stieltjes transforms of the c.d.f.s $\underline{F}^{\{y,H\}} \equiv (1-y)I_{[0,+\infty)} + yF^{\{y,H\}}$ and $\underline{F}^{\{y_p,H_p\}} \equiv (1-y_p)I_{[0,+\infty)} + y_pF^{\{y_p,H_p\}}$, respectively. Clearly, $\underline{F}^{\mathbf{B}_p} = (1-y_p)I_{[0,+\infty)} + y_pF^{\mathbf{B}_p}$ is the ESD of the matrix

$$\underline{\mathbf{B}}_p = \frac{1}{n}\left(\mathbf{x}_j^*T\mathbf{x}_k\right)_{j,k=1}^{n}.$$

Therefore, $\underline{F}^{\{y,H\}}$ and $\underline{m}$ are the LSD of $\underline{F}^{\mathbf{B}_p}$ and its Stieltjes transform and $\underline{F}^{\{y_p,H_p\}}$ and $m_{\underline{F}^{\{y_p,H_p\}}}$ are the corresponding versions with $y = y_p$ and $H = H_p$. Silverstein (1995) proved

Theorem 10. *If $H_p \to H$ and H is a proper probability distribution, then with probability 1, the ESD of $\mathbf{B_p}$ tends to a LSD for which the Stieltjes transform $\underline{m}$ is the unique solution to the equation*

$$z = -\frac{1}{\underline{m}(z)} + y \int \frac{t}{1 + t \cdot \underline{m}(z)} dH(t) \tag{3.8}$$

on the upper half plane $\underline{m} \in \mathbb{C}^+$.

When $H_p = 1_{[1,\infty)}$, that is $T_p = I_p$, a special of Theorem 10 is due to Marčenko and Pastur. For this special case, the LSD is called the MP law whose explicit form is given by

$$F_y'(x) = \begin{cases} \frac{1}{2\pi xy}\sqrt{(b-x)(x-a)} & \text{if } a < x < b, \\ 0 & \text{otherwise,} \end{cases} \tag{3.9}$$

where $a, b = (1 \mp \sqrt{y})^2$. If $y > 1$, there is a point mass $1 - 1/y$ at 0. The Stieltjes transform for the MP law is given by

$$\underline{m}(z) = -\frac{1 + z - y - \sqrt{(1 - z + y)^2 - 4z}}{2z}. \tag{3.10}$$

Next, we introduce the central limit theorem (CLT) of linear spectral statistics (LSS) due to Bai and Silverstein (2004). Suppose we are concerned with a parameter $\theta = \int f(x)dF(x)$. As an estimator of θ, one may employ the integral

$$\hat{\theta} = \int f(x)dF_n(x),$$

which will be called a LSS, where $F_n(x)$ is the ESD of the random matrix computed from data and $F(x)$ is the *limiting spectral distribution* (LSD) of $F_n \xrightarrow{D} F$.

Bai and Silverstein (2004) established the following theorem.

Theorem 11. *Assume:*

(a) *For each n, $x_{ij}^{(n)}$, $i \le p$, $j \le n$ are independent and for all n, i, j, they are identically distributed and $Ex_{11} = 0$, $E|x_{11}|^2 = 1$, $E|x_{11}^4| < \infty$;*
(b) *$y_p = p/n \to y$;*
(c) *T_p is $p \times p$ non-random Hermitian nonnegative definite with spectral norm bounded in p, with $F^{T_p} \xrightarrow{D} H$, a proper c.d.f. where D denotes the convergence in distribution.*

(d) *The function* $f_1, \ldots, f_k$ *are analytic on an open region of* $\mathbb{C}$ *containing the real interval*

$$\left[\liminf_p \lambda_{\min}^{T_p} \cdot I_{(0,1)}(y) \cdot (1-\sqrt{y})^2, \quad \limsup_p \lambda_{\max}^{T_p} \cdot (1+\sqrt{y})^2\right].$$

Then

(i) *the random vector*

$$\left(\int f_1(x) dG_p(x), \ldots, \int f_k(x) dG_p(x)\right) \tag{3.11}$$

forms a tight sequence in p*, where* $G_p(x) = p[F^{B_p}(x) - F^{\{y_p, H_p\}}(x)]$.

(ii) *If* x_{11} *is real and* $E(x_{11}^4) = 3$*, then* (3.11) *converges weakly to a Gaussian vector* $(X_{f_1}, \ldots, X_{f_k})$ *with means*

$$EX_f = -\frac{1}{2\pi i} \oint f(z) \frac{y \int \underline{m}(z)^3 \cdot t^2 \cdot (1 + t\underline{m}(z))^{-3} dH(t)}{\left[1 - y \int \underline{m}(z)^2 \cdot t^2 \cdot (1 + t\underline{m}(z))^{-2} dH(t)\right]^2} dz \tag{3.12}$$

and covariance function

$$\mathrm{Cov}(X_f, X_g) = -\frac{1}{2\pi^2} \oint\oint \frac{f(z_1) g(z_2)}{(\underline{m}(z_1) - \underline{m}(z_2))^2} \frac{d}{dz_1} \underline{m}(z_1) \frac{d}{dz_2} \underline{m}(z_2) dz_1 dz_2 \tag{3.13}$$

$(f, g \in \{f_1, \ldots, f_k\})$. *The contours in* (3.12) *and* (3.13) (*two in* (3.13)*, which we may assume to be nonoverlapping*) *are closed and are taken in the positive direction in the complex plane, each enclosing the support of* $F^{c,H}$.

(iii) *If* X_{11} *is complex with* $E(X_{11}^2)$ *and* $E(|X_{11}^4|) = 2$*, then* (*ii*) *also holds, except the means are zero and the covariance function is* $\frac{1}{2}$ *of the function given in* (3.13).

In the following, we consider the special case of $T_p = \mathbf{I}_{p \times p}$, that is,

$$\mathbf{S}_p = \mathbf{B}_p = \frac{1}{n} \sum_{k=1}^{n} \mathbf{X}_k \mathbf{X}_k^*;$$

then the ESD or LSD $H_p(t) = H(t)$ of F^{T_p} is a degenerate distribution in 1. Applying the theorem of Bai and Silverstein (2004), we obtain the following theorem.

Theorem 12. *If T_p is a $p \times p$ identity matrix $\mathbf{I}_{p\times p}$, then the results in Theorem 11 may be changed to*

$$\int f(x)dF^{y_p,H_p}(x) = \int_{a(y_p)}^{b(y_p)} \frac{f(x)}{2\pi y_p x}\sqrt{(b(y_p)-x)(x-a(y_p))}dx,$$

$$EX_f = \frac{f(a(y))+f(b(y))}{4} - \frac{1}{2\pi}\int_{a(y)}^{b(y)} \frac{f(x)}{\sqrt{4y-(x-1-y)^2}}dx$$

and

$$\mathrm{Cov}(X_f, X_g) = -\frac{1}{2\pi^2}\oint\oint \frac{f(z(\underline{m}_1))\cdot g(z(\underline{m}_2))}{(\underline{m}_1-\underline{m}_2)^2}d\underline{m}_1 d\underline{m}_2,$$

where

$$y_p = p/n \to y \in (0,1),$$

$$z(\underline{m}_i) = -\frac{1}{\underline{m}_i} + \frac{y}{1+\underline{m}_i} \quad \mathit{for} \quad i = 1,2,$$

$$a(y_p) = (1-\sqrt{y_p})^2, \quad b(y_p) = (1+\sqrt{y_p})^2$$

and the $\underline{m}_1$, $\underline{m}_2$ contours, nonintersecting and both taken in the positive direction, enclose $1/(y_p-1)$ and -1.

3.3. *Testing based on RMT limiting CLT*

In this section, we present a new testing method for the hypothesis H_0 by renormalizing T_1 and T_2 using the CLT for LSS of large sample covariance matrices. We would like to point out to the reader that our new approach applies to both cases of large dimension and small dimension, provided $p \geq 5$. From simulation comparisons, one can see that, when classical methods work well, our new approach is not only as good as the classical ones, but also performs well when the classical methods fail.

Based on the MP law (see Theorem 2.5 in Bai (1999)), we have the following lemma.

Lemma 13. *As $y_p = y/p \to y \in (0,1)$, with probability 1, we have*

$$\frac{\mathrm{tr}(S_n) - \log(|S_n|) - p}{p} \to d_2(y) \quad \mathit{under} \quad H_0,$$

where

$$d_2(y) = \int_a^b \frac{x - \log(x) - 1}{2\pi yx} \sqrt{(b(y) - x)(x - a(y))}dx$$
$$= 1 - \frac{y-1}{y} \log(1-y) < 0.$$

The proof is routine and omitted.

This limit theoretically confirms our findings that the classical methods of using T_1 and T_2 will lead to a very serious error in large-dimensional testing problems (3.1), that is, the Type I errors is almost 1. It suggests that one has to find a new normalization of the statistics T_1 and T_2 such that the hypothesis H_0 can be tested by the newly normalized versions of T_1 and T_2.

Applying Theorem 12 to T_1 and T_2, we have the following theorem.

Theorem 14. *When $y_p = p/n \to y \in (0,1)$, we have*

$$T_3 = \frac{\log(|S_n|) - p \cdot d_1(y_p) - \mu_1(y)}{\sigma_1(y)} \xrightarrow{D} N(0,1)$$
$$T_4 = \frac{\mathrm{tr}(S_n) - \log(|S_n|) - p - p \cdot d_2(y_p) - \mu_2(y_p)}{\sigma_2(y_p)} \xrightarrow{D} N(0,1)$$

where

$$d_1(y_p) = \frac{y_p - 1}{y_p} \log(1 - y_p) - 1,$$
$$\mu_1(y_p) = \frac{\log(1 - y_p)}{2},$$
$$\sigma_1^2(y_p) = -2\log(1 - y_p),$$
$$d_2(y_p) = 1 - \frac{y_p - 1}{y_p} \log(1 - y_p),$$
$$\mu_2(y_p) = -\frac{\log(1 - y_p)}{2},$$
$$\sigma_2^2(y_p) = -2\log(1 - y_p) - 2y_p.$$

The proof of the theorem is a simple application of of Theorem 12 and hence is omitted. We just present some simulation results to demonstrate how the new approach performs better than the original approaches.

3.4. *Simulation results*

In this section we present the simulation results for T_1, T_2, T_3 and T_4. We first investigate whether the Type I errors of the four methods can be controlled by the significance level. Then we investigate which of the four methods has the largest power. The purpose of this section is to show that the new hypothesis testing method T_4 provides a useful tool for both small and large dimensional problems (3.1)

$$H_0: \Sigma_{p\times p} = \mathbf{I}_{p\times p} \quad v.s. \quad H_1: \Sigma_{p\times p} \neq \mathbf{I}_{p\times p}.$$

Let us first give some detailed explanations. Firstly, we introduce the four simulated testing statistics with their respective approximations

$$\begin{cases} T_1 = \sqrt{\dfrac{n}{2p}} \cdot \log(|S_n|) \overset{D}{\sim} N(0,1), & \text{under } H_0, \\ T_2 = n \cdot (\mathrm{tr}(S_n) - \log(|S_n|) - p) \overset{D}{\sim} \chi^2_{p(p+1)/2}, & \text{under } H_0, \\ T_3 = \dfrac{\log(|S_n|) - p \cdot d_1(y_p) - \mu_1(y_p)}{\sigma_1(y_p)} \overset{D}{\sim} N(0,1), & \text{under } H_0, \\ T_4 = \dfrac{\mathrm{tr}(S_n) - \log(|S_n|) - p - p \cdot d_2(y_p) - \mu_2(y_p)}{\sigma_2(y_p)} \overset{D}{\sim} N(0,1), & \text{under } H_0. \end{cases}$$

Secondly, in order to illustrate detailed behaviors of the four statistics T_1–T_4, we not only use the two-sided rejection regions, but we also use the one-sided rejection regions in our simulation study.

$$\text{Method 1} \begin{cases} T_1 = \sqrt{\dfrac{n}{2p}} \cdot |\log(|S_n|)| \geq \alpha_{0.975} & \text{(two-sided)} \\ T_1 = \sqrt{\dfrac{n}{2p}} \cdot \log(|S_n|) \leq \alpha_{0.05} & \text{(reject left)} \\ T_1 = \sqrt{\dfrac{n}{2p}} \cdot \log(|S_n|) \geq \alpha_{0.95} & \text{(reject right)} \end{cases}$$

$$\text{Method 2} \begin{cases} T_2 = n \cdot (\mathrm{tr}(S_n) - \log(|S_n|) - p) \geq \beta_{0.975} \text{ or } \leq \beta_{0.025} & \text{(two sided)} \\ T_2 = n \cdot (\mathrm{tr}(S_n) - \log(|S_n|) - p) \leq \beta_{0.05} & \text{(reject left)} \\ T_2 = n \cdot (\mathrm{tr}(S_n) - \log(|S_n|) - p) \geq \beta_{0.95} & \text{(reject right)} \end{cases}$$

$$\text{Method 3}\begin{cases} T_3 = \dfrac{|\log(|S_n|) - p \cdot d_1(y_p) - \mu_1(y_p)|}{\sigma_1(y_p)} \geq \alpha_{0.975} & \text{(two-sided)} \\ T_3 = \dfrac{(\log(|S_n|) - p \cdot d_1(y_p) - \mu_1(y_p))}{\sigma_1(y_p)} \leq \alpha_{0.05} & \text{(reject left)} \\ T_3 = \dfrac{(\log(|S_n|) - p \cdot d_1(y_p) - \mu_1(y_p))}{\sigma_1(y_p)} \geq \alpha_{0.95} & \text{(reject right)} \end{cases}$$

$$\text{Method 4}\begin{cases} T_4 = \dfrac{|\operatorname{tr}(S_n) - \log(|S_n|) - p - p \cdot d_2(y_p) - \mu_2(y_p)|}{\sigma_2(y_p)} \geq \alpha_{0.975} \\ \qquad \text{(two-sided)} \\ T_4 = \dfrac{(\operatorname{tr}(S_n) - \log(|S_n|) - p - p \cdot d_2(y_p) - \mu_2(y_p))}{\sigma_2(y_p)} \leq \alpha_{0.05} \\ \qquad \text{(reject left)} \\ T_4 = \dfrac{(\operatorname{tr}(S_n) - \log(|S_n|) - p - p \cdot d_2(y_p) - \mu_2(y_p))}{\sigma_2(y_p)} \geq \alpha_{0.95} \\ \qquad \text{(reject right)} \end{cases}$$

where $\alpha_{0.975}$, $\alpha_{0.05}$ and $\alpha_{0.95}$ are the 97.5%, 5% and 95% quantiles of $N(0,1)$; $\beta_{0.975}$, $\beta_{0.05}$ and $\beta_{0.95}$ are the 97.5%, 5% and 95% quantiles of $\chi^2_{p(p+1)/2}$.

Thirdly, samples $X_1, \ldots, X_n$ are drawn from the population $N(0_p, \Sigma_{p\times p})$. To compute Type I errors, we draw samples $X_1, \ldots, X_n$ from $N(0_p, \mathbf{I}_{p\times p})$, and, to compute powers, we take samples $X_1, \ldots, X_n$ from $N(0_p, \mathbf{\Sigma}_{p\times p})$ where $\Sigma = (\sigma_{ij})_{p\times p}$

$$\sigma_{ij} = \begin{cases} 1, & i = j \\ 0.05, & i \neq j \end{cases}$$

for $i, j = 1, \ldots, p$. The sample size n is taken values 500 or 1000. The dimension of data p is taken values 5, 10, 50, 100 or 300. We also consider the case that the dimension p increases with the sample size n. The parameter setups are $(n,p) = (6000, 300)$, $(2000, 100)$, $(1000, 50)$, $(500, 25)$, $(250, 12)$ with $p/n = 0.05$.

The results of the 40 000 simulations are summarized in the following three tables.

Table 1. Type I errors and powers.

	Type I error			Power		
$(n = 500, p = 300)$	Two-sided	Reject left	Reject right	Two-sided	Reject left	Reject right
Method 1	1.0	1.0	0.0	1.0	1.0	0.0
Method 2	1.0	0.0	1.0	1.0	0.0	1.0
Method 3	0.0513	0.0508	0.0528	1.0	1.0	0.0
Method 4	0.0507	0.0521	0.0486	1.0	0.0	1.0
$(n = 500, p = 100)$	Two-sided	Reject left	Reject right	Two-sided	Reject left	Reject right
Method 1	1.0	1.0	0.0	1.0	1.0	0.0
Method 2	0.9523	0.0	0.9753	1.0	0.0	1.0
Method 3	0.0516	0.0514	0.0499	0.9969	1.0	0.0
Method 4	0.0516	0.0488	0.0521	1.0	0.0	1.0
$(n = 500, p = 50)$	Two-sided	Reject left	Reject right	Two-sided	Reject left	Reject right
Method 1	1.0	1.0	0.0	1.0	1.0	0.0
Method 2	0.1484	0.0064	0.2252	1.0	0.0	1.0
Method 3	0.0488	0.0471	0.0504	0.7850	0.8660	0.0
Method 4	0.0515	0.0494	0.0548	1.0	0.0	1.0
$(n = 500, p = 10)$	Two-sided	Reject left	Reject right	Two-sided	Reject left	Reject right
Method 1	0.0903	0.1406	0.0136	0.1712	0.2610	0.0023
Method 2	0.0546	0.0458	0.0538	0.8985	0.0	0.9391
Method 3	0.0507	0.0524	0.0489	0.0732	0.1169	0.0168
Method 4	0.0585	0.0441	0.0668	0.9252	0.0	0.9470
$(n = 500, p = 5)$	Two-sided	Reject left	Reject right	Two-sided	Reject left	Reject right
Method 1	0.0567	0.0777	0.0309	0.0651	0.1038	0.0190
Method 2	0.0506	0.0489	0.0511	0.4169	0.0014	0.5188
Method 3	0.0507	0.0517	0.0497	0.0502	0.0695	0.0331
Method 4	0.0625	0.0368	0.0807	0.5237	0.0007	0.5940

From the simulation results, one can see the following:

(1) Under all setups of (n, p), the simulated Type I errors of testing Methods 3 and 4 are close to the significance level $\alpha = 0.05$ while those of the testing Methods 1 and 2 are not. Moreover, when the ratio of dimension to sample size p/n is large, Type I errors of Methods 1 and 2 are close to 1.

Table 2. Type I errors and powers.

	Type I error			Power		
	($n = 1000, p = 300$)					
	Two-sided	Reject left	Reject right	Two-sided	Reject left	Reject right
Method 1	1.0	1.0	0.0	1.0	1.0	0.0
Method 2	1.0	0.0	1.0	1.0	0.0	1.0
Method 3	0.0496	0.0496	0.0493	1.0	1.0	0.0
Method 4	0.0512	0.0492	0.0499	1.0	0.0	1.0
	($n = 1000, p = 100$)					
	Two-sided	Reject left	Reject right	Two-sided	Reject left	Reject right
Method 1	1.0	1.0	0.0	1.0	1.0	0.0
Method 2	0.4221	0.0003	0.5473	1.0	0.0	1.0
Method 3	0.0508	0.0509	0.0515	1.0	1.0	0.0
Method 4	0.0522	0.0492	0.0535	1.0	0.0	1.0
	($n = 1000, p = 50$)					
	Two-sided	Reject left	Reject right	Two-sided	Reject left	Reject right
Method 1	0.9830	0.9915	0.0	1.0	1.0	0.0
Method 2	0.0778	0.0179	0.1166	1.0	0.0	1.0
Method 3	0.0471	0.0495	0.0499	0.9779	0.9886	0.0
Method 4	0.0524	0.0473	0.0575	1.0	0.0	1.0
	($n = 1000, p = 10$)					
	Two-sided	Reject left	Reject right	Two-sided	Reject left	Reject right
Method 1	0.0666	0.1067	0.0209	0.1801	0.2623	0.0037
Method 2	0.0532	0.0470	0.0517	0.9994	0.0	0.9995
Method 3	0.0506	0.0498	0.0504	0.0969	0.1591	0.0116
Method 4	0.0582	0.0440	0.0669	0.9994	0.0	0.9996
	($n = 1000, p = 5$)					
	Two-sided	Reject left	Reject right	Two-sided	Reject left	Reject right
Method 1	0.0530	0.0696	0.0360	0.0664	0.1040	0.0203
Method 2	0.0510	0.0491	0.0498	0.8086	0.0	0.8736
Method 3	0.0508	0.0530	0.0494	0.0542	0.0784	0.0288
Method 4	0.0622	0.0356	0.0790	0.8780	0.0	0.9114

Furthermore, when the ratio $p/n \to y \in (0, 1)$, even if y is very small, Type I errors of testing Methods 1 and 2 still tend to 1 as the sample size is becoming large.

(2) Under all choices of (n, p), powers of testing Methods 2 and 4 are much higher than those of testing Methods 1 and 3, respectively. Moreover, almost all powers of testing Method 4 are higher than others.

Table 3. Type I errors and powers.

	Type I error			Power		
$(n = 6000, p = 300)$						
	Two-sided	Reject left	Reject right	Two-sided	Reject left	Reject right
Method 1	1.0	1.0	0.0	1.0	1.0	0.0
Method 2	0.7186	0.0	0.8131	1.0	0.0	1.0
Method 3	0.0476	0.0465	0.0469	1.0	1.0	0.0
Method 4	0.0505	0.0525	0.0466	1.0	0.0	1.0
$(n = 2000, p = 100)$						
	Two-sided	Reject left	Reject right	Two-sided	Reject left	Reject right
Method 1	1.0	1.0	0.0	1.0	1.0	0.0
Method 2	0.1366	0.0062	0.2144	1.0	0.0	1.0
Method 3	0.0501	0.0506	0.0515	1.0	1.0	0.0
Method 4	0.0525	0.0505	0.0531	1.0	0.0	1.0
$(n = 1000, p = 50)$						
	Two-sided	Reject left	Reject right	Two-sided	Reject left	Reject right
Method 1	0.9830	0.9915	0.0	1.0	1.0	0.0
Method 2	0.0778	0.0179	0.1166	1.0	0.0	1.0
Method 3	0.0471	0.0495	0.0499	0.9779	0.9886	0.0
Method 4	0.0524	0.0473	0.0575	1.0	0.0	1.0
$(n = 500, p = 25)$						
	Two-sided	Reject left	Reject right	Two-sided	Reject left	Reject right
Method 1	0.5511	0.6653	0.0	0.9338	0.9656	0.0
Method 2	0.0817	0.0313	0.0765	1.0	0.0	1.0
Method 3	0.0518	0.0539	0.0506	0.2824	0.3948	0.0013
Method 4	0.0552	0.0472	0.0558	1.0	0.0	1.0
$(n = 250, p = 12)$						
	Two-sided	Reject left	Reject right	Two-sided	Reject left	Reject right
Method 1	0.1835	0.2729	0.0033	0.3040	0.4151	0.0006
Method 2	0.0612	0.0442	0.0606	0.6129	0.0002	0.7141
Method 3	0.0483	0.0499	0.0486	0.0670	0.1089	0.0183
Method 4	0.0574	0.0507	0.0617	0.6369	0.0003	0.7192

(3) Comparing the Type I errors and powers for all choices of (n, p), the testing Method 4 has better Type I errors and higher powers. Although Method 2 has higher powers, its Type I errors are almost 1. Although Method 3 has lower Type I errors, its powers are lower than those of Method 4.

In conclusion, our simulation results show that T_1 and T_2 are inapplicable for large-dimensional problems or small-dimensional problems whose sample size is large. Although both statistics, T_3 and T_4, can be applied to the large-dimensional problem (3.1), T_4 is better than T_3 from the view point of powers under the same significance level. It shows that T_4 provides a robust test for both, large-dimensional or small-dimensional problems.

4. Conclusions

In this paper, both theoretically and by simulation, we have shown that classical approaches to hypothesis testing do not apply to large-dimensional problems and that the newly proposed methods perform much better than the classical ones. It is interesting that the new methods do not perform much worse than the classical methods for small dimensional cases. Therefore, we would strongly recommend the new approaches even for moderately large dimensional cases provided that $p \geq 4$ or 5, REGARDLESS of the ratio between dimension and data size.

We would also like to emphasize that the large dimension of data may cause low efficiency of classical inference methods. In such cases, we would strongly recommend non-exact procedures with high efficiency rather than those classical ones with low efficiency, such as Dempster's NET and Bai and Saranadasa's ANT.

Acknowledgment

The authors of this chapter would like to express their thanks to Dr. Adrian Roellin for his careful proofreading of the chapter and valuable comments.

References

1. G. J. Babu and Z. D. Bai, Edgeworth expansions of a function of sample means under minimal moment conditions and partial Cramer's conditions, *Sankhya Ser. A* **55** (1993) 244–258.
2. Z. D. Bai, Convergence rate of expected spectral distributions of large random matrices. Part I. Wigner matrices, *Ann. Probab.* **21** (1993) 625–648.
3. Z. D. Bai, Convergence rate of expected spectral distributions of large random matrices. Part II. Sample covariance matrices, *Ann. Probab.* **21** (1993) 649–672.
4. Z. D. Bai, Methodologies in spectral analysis of large dimensional random matrices. A review, *Statistica Sinica* **9** (1999) 611–677.
5. Z. D. Bai and C. R. Rao, Edgeworth expansion of a function of sample means, *Ann. Statist.* **19** (1991) 1295–1315.

6. Z. D. Bai and J. W. Silverstein, No eigenvalues outside the support of the limiting spectral distribution of large dimensional sample covariance matrices, *Ann. Probab.* **26** (1998) 316–345.
7. Z. D. Bai and J. W. Silverstein, CLT for linear spectral statistics of large-dimensional sample covariance matrices, *Ann. Probab.* **32** (2004) 553–605.
8. Z. D. Bai and Y. Q. Yin, Limit of the smallest eigenvalue of large dimensional sample covariance matrix, *Ann. Probab.* **21** (1993) 1275–1294.
9. Z. D. Bai, Y. Q. Yin and P. R. Krishnaiah, On the limiting empirical distribution function of the eigenvalues of a multivariate F-matrix, *Theory Probab. Appl.* **32** (1987) 490–500.
10. J. H. Chung and D. A. S. Fraser, Randomization tests for a multivariate two-sample problem, *J. Amer. Statist. Assoc.* **53** (1958) 729–735.
11. A. P. Dempster, A high dimensional two sample significance test, *Ann. Math. Statist.* **29** (1958) 995–1010.
12. A. P. Dempster, A significance test for the separation of two highly multivariate small samples, *Biometrics* **16** (1960) 41–50.
13. V. A. Marčenko and L. A. Pastur, Distribution of eigenvalues for some sets of random matrices, *Math. USSR-Sb* **1** (1967) 457–483.
14. H. Saranadasa, Asymptotic expansion of the missclassification probabilities of D- and A-criteria for discrimination from two high dimensional populations using the theory of large dimensional random matrices, *J. Multivariate Anal.* **46** (1993) 154–174.
15. Y. Q. Yin, Z. D. Bai and P. R. Krishnaiah, Limiting behavior of the eigenvalues of a multivariate F-matrix, *J. Multivariate Anal.* **13** (1983) 508–516.
16. Y. Q. Yin, Z. D. Bai and P. R. Krishnaiah, On the limit of the largest eigenvalue of the large dimensional sample covariance matrix, *Probab. Theory Related Fields* **78** (1988) 509–521.

THE η AND SHANNON TRANSFORMS: A BRIDGE BETWEEN RANDOM MATRICES AND WIRELESS COMMUNICATIONS

Antonia M. Tulino

Dip. di Ing. Elettronica e delle Telecomunicazioni
Universita' degli Studi di Napoli, "Fedrico I
Via Claudio 21, Napoli, Italy
E-mail: atulino@princeton.edu

The landmark contributions to the theory of random matrices of Wishart (1928), Wigner (1955), and Marčenko-Pastur (1967), were motivated to a large extent by their applications. In this paper, we report on two transforms motivated by the application of random matrices to various problems in the information theory of noisy communication channels: η and Shannon transforms. Originally introduced in [1, 2], their applications to random matrix theory and engineering applications have been developed in [3]. In this paper, we give a summary of their main properties and applications in random matrix theory.

1. Introduction

The first studies of random matrices stemmed from the multivariate statistical analysis at the end of the 1920s, primarily with the work of Wishart (1928) on fixed-size matrices with Gaussian entries. After a slow start, the subject gained prominence when Wigner introduced the concept of statistical distribution of nuclear energy levels in 1950. In the past half century, classical random matrix theory has been developed, widely and deeply, into a huge body, effectively used in many branches of physics and mathematics. Of late, random matrices have attracted great interest in the engineering community because of their applications in the context of information theory and signal processing, which include among others: wireless communications channels, learning and neural networks, capacity of ad hoc networks, direction of arrival estimation in sensor arrays, etc.

The earliest applications to wireless communication were the pioneering works of Foschini and Telatar in the mid-90s on characterizing the capacity of multi-antenna channels. With works like [4–6] which, initially, called attention to the effectiveness of asymptotic random matrix theory in wireless communication theory, interest in the study of random matrices began and the singular-value densities of random matrices and their asymptotics, as the matrix size tends to infinity, became an active research area in information/communication theory. In the last few years a considerable body of results on the fundamental information-theoretic limits of various wireless communication channels that makes substantial use of asymptotic random matrix theory, has emerged in the communications and information theory literature. For an extended survey on contributions on this results see [3].

In the same way that the original contributions of Wishart and Wigner were motivated by their applications, such is also the driving force behind the efforts by information-theoreticians and engineers. The Shannon and the η transforms, introduced for the first time in [1,2], are prime examples: these transforms which were motivated by the application of random matrix theory to various problems in the information theory of noisy communication channels [3], characterize the spectrum of a random matrix while providing direct engineering insight.

In this paper, using the η and Shannon transforms of the singular-value distributions of large dimensional random matrices, we characterize for both ergodic and non-ergodic regime the fundamental limits of a general class of noisy multi-input multi-output (MIMO) wireless channels which are characterized by random matrices that admit various statistical descriptions depending on the actual application. For these channels, a number of examples and asymptotic closed-form expressions of their fundamental limits are provided. For both the ergodic and non-ergodic regimes, we illustrate the power of random matrix results in the derivation of the fundamental limits of wireless channels and we show the applicability of our results to real-world problems, where the asymptotic behaviors are shown to be excellent approximations of the behavior of actual systems with very modest numbers of antennas.

2. Wireless Communication Channels

A typical wireless communication channel is described by the usual linear vector memoryless channel:

$$\mathbf{y} = \mathbf{H}\mathbf{x} + \mathbf{n} \tag{2.1}$$

where $\mathbf{x}$ is the K-dimensional vector of the signal input, $\mathbf{y}$ is the N-dimensional vector of the signal output, and the N-dimensional vector $\mathbf{n}$ is the additive Gaussian noise, whose components are independent complex Gaussian random variables with zero mean and independent real and imaginary parts with the same variance $\sigma^2/2$ (i.e., circularly distributed). $\mathbf{H}$, in turn, is the $N \times K$ complex random matrix describing the channel.

Model (2.1) encompasses a variety of channels of interest in wireless communications such as multi-access channels, linear channels with frequency-selective and/or frequency-dispersive fading, multidimensional channels (multi-sensor reception, multi-cellular system with cooperative detection, etc), crosstalk in digital subscriber lines, signal space diversity, etc. In each of these cases, N, K and $\mathbf{H}$ take different meanings. For example, K and N may indicate the number of transmit and receive antennas while $\mathbf{H}$ describes the fading between each pair of transmit and receive antennas, or the spreading gain and the number of users while $\mathbf{H}$ the signature matrix, or they may both represent time/frequency slots while $\mathbf{H}$ the tone matrix.

In Section 5 we detail some of the more representative wireless channels described by (2.1) that capture various features of interest in wireless communications and we demonstrate how random matrix results — along with the η and Shannon transforms — have been used to characterize the fundamental limits of the various channels that arise in wireless communications.

3. Why Asymptotic Random Matrix Theory?

In section we illustrate the role of random matrices and their singular values in wireless communication through the derivation of some key performance measures, which are determined by the distribution of the singular values of the channel matrix.

The empirical cumulative distribution function (c.d.f) of the eigenvalues (also referred to as the empirical spectral distribution (ESD)) of an $N \times N$ Hermitian matrix $\mathbf{A}$ is defined as

$$\mathsf{F}^N_{\mathbf{A}}(x) = \frac{1}{N} \sum_{i=1}^{N} 1\{\lambda_i(\mathbf{A}) \leq x\} \tag{3.1}$$

where $\lambda_1(\mathbf{A}), \ldots, \lambda_N(\mathbf{A})$ are the eigenvalues of $\mathbf{A}$ and $1\{\cdot\}$ is the indicator function. If $\mathsf{F}^N_{\mathbf{A}}(\cdot)$ converges almost surely (a.s) as $N \to \infty$, then the corresponding limit (asymptotic ESD) is denoted by $\mathsf{F}_{\mathbf{A}}(\cdot)$.

The first performance measure that we are going to consider is the mutual information. The mutual information, first introduced by Shannon in 1948, determines the maximum amount of data per unit bandwidth (in bits/s/Hz) that can be transmitted reliably over a specific channel realization $\mathbf{H}$. If the channel is known by the receiver, and the input $\mathbf{x}$ is Gaussian with independent and identically distributed (i.i.d.) entries, the normalized mutual information in (2.1) conditioned on $\mathbf{H}$ is given by [7,8]

$$\begin{aligned}
\mathcal{I}(\mathsf{SNR}) &= \frac{1}{N} I(\mathbf{x};\mathbf{y}|\mathbf{H}) && (3.2)\\
&= \frac{1}{N} \log\det\left(\mathbf{I} + \mathsf{SNR}\,\mathbf{H}\mathbf{H}^\dagger\right) && (3.3)\\
&= \frac{1}{N} \sum_{i=1}^{N} \log\left(1 + \mathsf{SNR}\,\lambda_i(\mathbf{H}\mathbf{H}^\dagger)\right) && (3.4)\\
&= \int_0^\infty \log\left(1 + \mathsf{SNR}\,x\right) d\mathsf{F}^N_{\mathbf{H}\mathbf{H}^\dagger}(x) && (3.5)
\end{aligned}$$

with the transmitted signal-to-noise ratio (SNR)

$$\mathsf{SNR} = \frac{N\mathbb{E}[||\mathbf{x}||^2]}{K\mathbb{E}[||\mathbf{n}||^2]}, \tag{3.6}$$

and $\lambda_i(\mathbf{H}\mathbf{\Phi}\mathbf{H}^\dagger)$ equal to the ith squared singular value of $\mathbf{H}$.

If the channel is known at the receiver and its variation over time is stationary and ergodic, then the expectation of (3.2) over the distribution of $\mathbf{H}$ is the ergodic mutual information (normalized to the number of receive antennas or the number of degrees of freedom per symbol in the CDMA channel).

For $\mathsf{SNR} \to \infty$, a regime of interest in short-range applications, the normalized mutual information admits the following affine expansion [9,10]

$$\mathcal{I}(\mathsf{SNR}) = \mathcal{S}_\infty \left(\log \mathsf{SNR} + \mathcal{L}_\infty\right) + o(1) \tag{3.7}$$

where the key measures are the *high-SNR slope*

$$\mathcal{S}_\infty = \lim_{\mathsf{SNR}\to\infty} \frac{\mathcal{I}(\mathsf{SNR})}{\log \mathsf{SNR}} \tag{3.8}$$

which for most channels gives $\mathcal{S}_\infty = \min\left\{\frac{K}{N}, 1\right\}$, and the *power offset*

$$\mathcal{L}_\infty = \lim_{\mathsf{SNR}\to\infty} \log \mathsf{SNR} - \frac{\mathcal{I}(\mathsf{SNR})}{\mathcal{S}_\infty} \tag{3.9}$$

which essentially boils down to $\log\det(\mathbf{H}\mathbf{H}^\dagger)$ or $\log\det(\mathbf{H}^\dagger\mathbf{H})$ depending on whether $K > N$ or $K < N$.

Another important performance measure for (2.1) is the minimum mean-square-error (MMSE) achieved by a linear receiver, which determines the maximum achievable output signal-to-interference-and-noise ratio (SINR). For an i.i.d. input, the arithmetic mean over the users (or transmit antennas) of the MMSE is given, as a function of $\mathbf{H}$, by [4]

$$\mathsf{MMSE}(\mathsf{SNR}) = \frac{1}{K} \min_{\mathbf{M} \in C^{K \times N}} \mathbb{E}[||\mathbf{x} - \mathbf{M}\mathbf{y}||^2] \tag{3.10}$$

$$= \frac{1}{K} \operatorname{tr}\left\{ \left(\mathbf{I} + \mathsf{SNR}\, \mathbf{H}^\dagger \mathbf{H}\right)^{-1} \right\} \tag{3.11}$$

$$= \frac{1}{K} \sum_{i=1}^{K} \frac{1}{1 + \mathsf{SNR}\, \lambda_i(\mathbf{H}^\dagger \mathbf{H})} \tag{3.12}$$

$$= \int_0^\infty \frac{1}{1 + \mathsf{SNR}\, x} d\mathsf{F}^K_{\mathbf{H}^\dagger \mathbf{H}}(x)$$

$$= \frac{N}{K} \int_0^\infty \frac{1}{1 + \mathsf{SNR}\, x} d\mathsf{F}^N_{\mathbf{H}\mathbf{H}^\dagger}(x) - \frac{N-K}{K} \tag{3.13}$$

where the expectation in (3.10) is over $\mathbf{x}$ and $\mathbf{n}$ while (3.13) follows from

$$N\mathsf{F}^N_{\mathbf{H}\mathbf{H}^\dagger}(x) - Nu(x) = K\mathsf{F}^K_{\mathbf{H}^\dagger \mathbf{H}}(x) - Ku(x) \tag{3.14}$$

where $u(x)$ is the unit-step function ($u(x) = 0,\ x \le 0;\ u(x) = 1,\ x > 0$). Note, incidentally, that both performance measures as a function of SNR are coupled through

$$\frac{d}{d\,\mathsf{SNR}} \log_e \det\left(\mathbf{I} + \mathsf{SNR}\, \mathbf{H}\mathbf{H}^\dagger\right) = \frac{K - \operatorname{tr}\left\{ \left(\mathbf{I} + \mathsf{SNR}\, \mathbf{H}^\dagger \mathbf{H}\right)^{-1} \right\}}{\mathsf{SNR}}.$$

As we see in (3.5) and (3.13), both fundamental performance measures (mutual information and MMSE) are dictated by the distribution of the empirical (squared) singular value distribution of the random channel matrix. It is thus of paramount importance, in order to evaluate these — and other — performance measures, to be able to express this empirical distribution. Since $\mathsf{F}^N_{\mathbf{H}\mathbf{H}^\dagger}$ clearly depends on the specific realization of $\mathbf{H}$, so do (3.2) and (3.10) above. In terms of engineering insight, however, it is crucial to obtain expressions for the performance measures that do not depend on the single matrix realization, to which end two approaches are possible:

- To study the average behavior[a] by taking an expectation of the performance measures over **H**, which requires assigning a probabilistic structure to it.
- The second approach is to consider an operative regime where the performance measures (3.2) and (3.10) do not depend on the specific choice of signatures.

Asymptotic analysis (in the sense of large dimensional systems, i.e $K, N \to \infty$ with $\frac{K}{N} \to \beta$) is where both these approaches meet. First, the computation of the average performance measures simplifies as the dimensions grow to infinity. Second, the asymptotic regime turn out to be the operative regime where the dependencies of (3.2) and (3.10) on the realization of **H** disappear. Specifically, in most of the cases, asymptotic random matrix theory guarantees that as the dimensions of **H** go to infinity but their ratio is kept constant, its empirical singular-value distribution displays the following properties, which are key to the applicability to wireless communication problems:

- Insensitivity of the asymptotic eigenvalue distribution to the probability density function of the random matrix entries.
- An "ergodic" nature in the sense that — with probability one — the eigenvalue histogram of any matrix realization converges almost surely to the asymptotic eigenvalue distribution.
- Fast convergence rate of the empirical singular-value distribution to its asymptotic limit [11,12], which implies that that even for small values of the parameters, the asymptotic results come close to the finite-parameter results (cf. Fig. 1).

All these properties are very attractive in terms of analysis but are also of paramount importance at the design level. In fact:

- The ergodicity enables the design of a receiver that, when optimized in the asymptotic regime, has a structure depending weakly or even not depending at all of the specific realization of **H**. As a consequence, less *a priori* knowledge and a lower level of complexity are required (see [3] and references therein).

[a]It is worth emphasizing that, in many cases, resorting to the expected value of the mutual information is motivated by the stronger consideration that: in problems such as aperiodic DS-CDMA or multi-antenna with an ergodic channel, it is precisely the expected capacity that has real operational meaning.

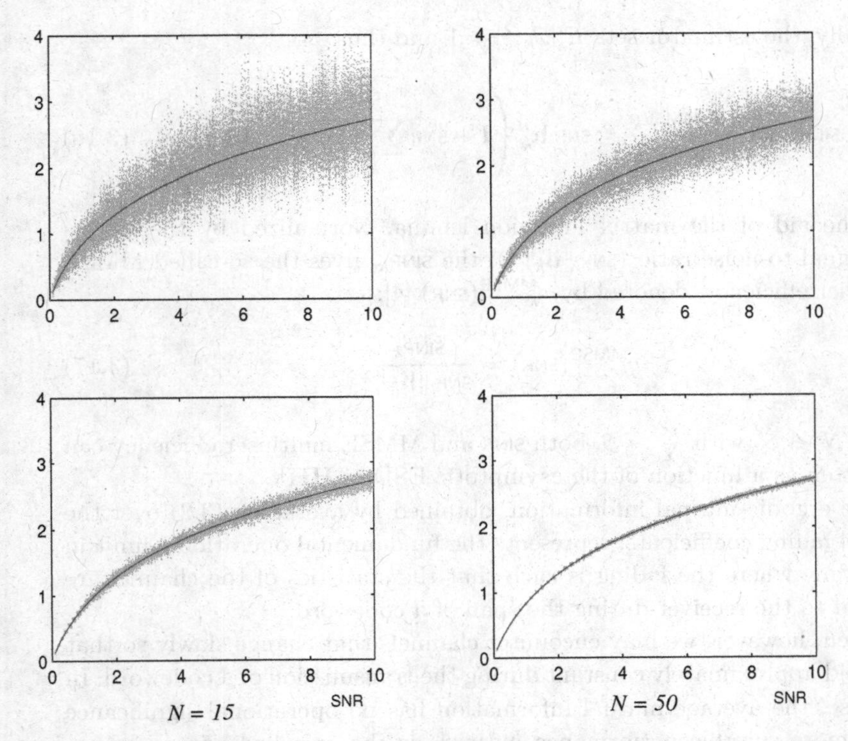

Fig. 1. Several realizations of the left hand side of (3.3) are compared to the asymptotic limit in the right hand side of (4.10) in the case of $\beta = 1$ for $N = 3, 5, 15, 50$.

- The fast convergence ensures that the performance of the asymptotically designed receiver operating for very small values of the system dimensions, is very close to that of the optimized receiver.
- Finally, the insensitivity property — along with the fast convergence — leads to receiver structures that, for finite dimensionality, are very robust to the probability distribution of $\mathbf{H}$. Examples are the cases of DS-CDMA subject to fading or single-user multiple-antennas link, where the results do not depend on the fading statistics.

As already mentioned, closely related to the MMSE is the signal-to-interference-to noise ratio, SINR, achieved at the output of a linear MMSE receiver. Denote by $\hat{x}_k$ the MMSE estimate of the kth component of $\mathbf{x}$ and by MMSE_k the corresponding MMSE, such SINR for the kth component is:

$$\mathsf{SINR}_k = \frac{E[|\hat{x}_k|^2] - \mathsf{MMSE}_k}{\mathsf{MMSE}_k}. \tag{3.15}$$

Typically, the estimator sets $E[|\hat{x}_k|^2] = 1$ and thus

$$\mathsf{SINR}_k = \frac{1 - \mathsf{MMSE}_k}{\mathsf{MMSE}_k} = \mathsf{SNR}\, \mathbf{h}_k^\dagger \left(\mathbf{I} + \mathsf{SNR} \sum_{j \neq k} \mathbf{h}_j \mathbf{h}_j^\dagger \right)^{-1} \mathbf{h}_k. \tag{3.16}$$

with the aid of the matrix inversion lemma. Normalized by the single-user signal-to-noise ratio ($\mathsf{SNR}\, \|\mathbf{h}_k\|^2$), the SINR_k gives the so-called MMSE multiuser efficiency, denoted by $\eta_k^{\mathsf{MMSE}}(\mathsf{SNR})$ [4]:

$$\eta_k^{\mathsf{MMSE}}(\mathsf{SNR}) = \frac{\mathsf{SINR}_k}{\mathsf{SNR}\, \|\mathbf{h}_k\|^2}. \tag{3.17}$$

For $K, N \to \infty$ with $\frac{K}{N} \to \beta$, both SINR and MMSE multiuser efficiency can be written as a function of the asymptotic ESD of $\mathbf{H}\mathbf{H}^\dagger$.

The ergodic mutual information, obtained by averaging (3.2) over the channel fading coefficients, represents the fundamental operational limit in the regime where the fading is such that the statistics of the channel are revealed to the receiver during the span of a codeword.

Often, however, we may encounter channels that change slowly so that $\mathbf{H}$ is held approximately constant during the transmission of a codeword. In this case, the average mutual information has no operational significance and a more suitable performance measure is the so-called outage capacity [13] (cumulative distribution of the mutual information), which coincides with the classical Shannon-theoretic notion of ϵ-capacity [14], namely the maximal rate for which block error probability ϵ is attainable. Under certain conditions, the outage capacity can be obtained through the probability that the transmission rate R exceeds the input-output mutual information (conditioned on the channel realization) [15,16,13]. Thus, given a rate R an outage occurs when the random variable

$$I = \log\det(\mathbf{I} + \mathsf{SNR}\, \mathbf{H}\mathbf{H}^\dagger) \tag{3.18}$$

whose distribution is induced by $\mathbf{H}$, falls below R. A central result in random matrix theory derived by Bai and Silverstein (2004) [17] establishes a law of large numbers and a central limit theorem for linear statistics of a suitable class of Hermitian random matrices. Using this result, in Section 4 the asymptotic normality of the unnormalized mutual information (3.18) is proved for arbitrary signal-to-noise ratios and fading distributions, allowing for correlation between either transmit or receive antennas.

4. η and Shannon Transforms: Theory and Applications

Motivated by the intuition drawn from various applications of random matrices to problems in the information theory of noisy communication channels, η and Shannon transforms, which are very related to a more classical transform in random matrix theory, the Stieltjes transform [18], turn out to be quite helpful at clarifying the exposition as well as the statement of many results. In particular, the η transform leads to compact definitions of other transforms used in random matrix theory such as the R and S transforms [3].

Definition 4.1. Given an $N \times N$ nonnegative definite random matrix $\mathbf{A}$ whose ESD converges a.s., its η transform is

$$\eta_{\mathbf{A}}(\gamma) = \mathbb{E}\left[\frac{1}{1+\gamma X}\right] \tag{4.1}$$

while its Shannon transform is defined as

$$\mathcal{V}_{\mathbf{A}}(\gamma) = \mathbb{E}[\log(1+\gamma X)] \tag{4.2}$$

where X is a nonnegative random variable whose distribution is the asymptotic ESD of $\mathbf{A}$ while γ is a nonnegative real number.

Then, $\eta_{\mathbf{A}}(\gamma)$ can be regarded as a generating function for the asymptotic moments of $\mathbf{A}$ [3]. Furthermore from the definition $0 < \eta_X(\gamma) \leq 1$.

For notational convenience, we refer to the transform of a matrix and the transform of its asymptotic ESD interchangeably.

Lemma 4.2. *For any $N \times K$ matrix $\mathbf{A}$ and $K \times N$ matrix $\mathbf{B}$ such that $\mathbf{AB}$ is nonnegative definite, for $K, N \to \infty$ with $\frac{K}{N} \to \beta$, if the spectra converge,*

$$\eta_{\mathbf{AB}}(\gamma) = 1 - \beta + \beta\eta_{\mathbf{BA}}(\gamma). \tag{4.3}$$

As it turns out, the Shannon and η transforms are intimately related to each other and to the classical Stieltjes transform:

$$\frac{\gamma}{\log e}\frac{d}{d\gamma}\mathcal{V}_{\mathbf{A}}(\gamma) = 1 - \frac{1}{\gamma}\mathcal{S}_{\mathbf{A}}\left(-\frac{1}{\gamma}\right) = 1 - \eta_{\mathbf{A}}(\gamma)$$

where $\mathbf{A}$ is $N \times N$ Hermitian matrix whose ESD converges a.s. to $F_{\mathbf{A}}(\cdot)$ and $S_{\mathbf{A}}(\cdot)$ is its Stieltjes transform defined as [18]:

$$S_{\mathbf{A}}(z) = \int \frac{1}{\lambda - z} dF_{\mathbf{A}}(\lambda). \tag{4.4}$$

Before introducing the η and Shannon transforms of various random matrices, some justification for their relevance to wireless communications is in order. The rationale for introducing the η and Shannon transforms can be succinctly explained by considering a hypothetical wireless communication channel where the random channel matrix $\mathbf{H}$ in (2.1) is such that, as $K, N \to \infty$ with $\frac{K}{N} \to \beta$, the ESD of $\mathbf{HH}^\dagger$ converges a.s. to a nonrandom limit. Based on Definition 4.1 we immediately recognize from (3.5) and (3.13) that for an i.i.d. Gaussian input $\mathbf{x}$, as $K, N \to \infty$ with $\frac{K}{N} \to \beta$ the normalized mutual information and the MMSE of (2.1) are related to η and Shannon transform of $\mathbf{HH}^\dagger$ by the following relationships:

$$\mathcal{I}(\mathsf{SNR}) \to \mathcal{V}_{\mathbf{HH}^\dagger}(\mathsf{SNR}) \tag{4.5}$$

$$\mathsf{MMSE}(\mathsf{SNR}) \to \eta_{\mathbf{H}^\dagger\mathbf{H}}(\mathsf{SNR}) \tag{4.6}$$

$$= 1 - \frac{1 - \eta_{\mathbf{HH}^\dagger}(\mathsf{SNR})}{\beta} \tag{4.7}$$

where (4.7) follows from (3.14). It is thus of vital interest in information-theoretic and signal-processing analysis of the wireless communication channels of contemporary interest, the evaluation of the η and Shannon transforms of the various random (channel) matrices that arise in the linear model (2.1).

A classical result in random matrix theory states that

Theorem 4.3 ([5]). *If the entries of* $\mathbf{H}$ *are zero-mean i.i.d. with variance* $\frac{1}{N}$, *as* $K, N \to \infty$ *with* $\frac{K}{N} \to \beta$, *the ESD of* $\mathbf{HH}^\dagger$ *converges a.s. to the Marčenko-Pastur law whose density function is*

$$\tilde{f}_\beta(x) = (1-\beta)^+ \, \delta(x) + \frac{\sqrt{(x-a)^+(b-x)^+}}{2\pi x} \tag{4.8}$$

where

$$a = (1-\sqrt{\beta})^2 \qquad\qquad b = (1+\sqrt{\beta})^2.$$

For this simple statistical structure of $\mathbf{H}$, the η and Shannon transforms admit the following nice and compact closed-form expressions:

Theorem 4.4 ([5]). *The* η *and Shannon transforms of the Marčenko-Pastur law, whose density function is (4.8), are*

$$\eta_{\mathbf{HH}^\dagger}(\gamma) = 1 - \frac{\mathcal{F}(\gamma,\beta)}{4\,\gamma} \tag{4.9}$$

and

$$\mathcal{V}_{\mathbf{HH}^\dagger}(\gamma) = \beta \log\left(1+\gamma-\frac{1}{4}\mathcal{F}(\gamma,\beta)\right) + \log\left(1+\gamma\beta-\frac{1}{4}\mathcal{F}(\gamma,\beta)\right) - \frac{\log e}{4\,\gamma}\mathcal{F}(\gamma,\beta) \tag{4.10}$$

with

$$\mathcal{F}(x,z) = \left(\sqrt{x(1+\sqrt{z})^2+1} - \sqrt{x(1-\sqrt{z})^2+1}\right)^2 .$$

However, as is well known since the work of Marčenko and Pastur [19], it is rare the case that the limiting empirical distribution of the squared singular values of random matrices (whose aspect ratio converges to a constant) admit closed-form expressions. Rather, [19] showed a very general result where the characterization of the solution is accomplished through a fixed-point equation involving the Stieltjes transform. Later this result has been strengthened in [20]. Consistent with our emphasis, this result is formulated in terms of the η transform rather than the Stieltjes transform used in [20] as follows:

Theorem 4.5 ([19, 20]). *Let* $\mathbf{S}$ *be an* $N \times K$ *matrix whose entries are i.i.d. complex random variables with zero-mean and variance* $\frac{1}{N}$. *Let* $\mathbf{T}$ *be a* $K \times K$ *real diagonal random matrix whose empirical eigenvalue distribution converges a.s. to a nonrandom limit. Let* $\mathbf{W}_0$ *be an* $N \times N$ *Hermitian complex random matrix with empirical eigenvalue distribution converging a.s. to a nonrandom distribution. If* $\mathbf{H}$, $\mathbf{T}$, *and* $\mathbf{W}_0$ *are independent, the empirical eigenvalue distribution of*

$$\mathbf{W} = \mathbf{W}_0 + \mathbf{STS}^\dagger \tag{4.11}$$

converges, as K, $N \to \infty$ *with* $\frac{K}{N} \to \beta$, *a.s. to a nonrandom limiting distribution whose* η *transform is the solution of the following pair of equations:*

$$\gamma\,\eta = \varphi\,\eta_0\,(\varphi) \tag{4.12}$$

$$\eta = \eta_0\,(\varphi) - \beta\,(1 - \eta_{\mathbf{T}}(\gamma\,\eta)) \tag{4.13}$$

with η_0 *and* $\eta_{\mathbf{T}}$ *the* η *transforms of* $\mathbf{W}_0$ *and* $\mathbf{T}$ *respectively.*

In the following we give some of the more representative results on the η and Shannon transform, where the Shannon and η transforms lead to particularly simple solutions for the limiting empirical distribution of the squared singular values of random matrices with either dependent or independent entries.

Theorem 4.6 ([3]). *Let* $\mathbf{S}$ *be an* $N \times K$ *complex random matrix whose entries are i.i.d. with variance* $\frac{1}{N}$. *Let* $\mathbf{T}$ *be a* $K \times K$ *nonnegative definite random matrix, whose ESD converges a.s. to a nonrandom distribution. The*

ESD of $\mathbf{STS}^\dagger$ *converges a.s., as* $K, N \to \infty$ *with* $\frac{K}{N} \to \beta$, *to a distribution whose* η *transform satisfies*

$$\beta = \frac{1-\eta}{1-\eta_{\mathbf{T}}(\gamma\eta)} \tag{4.14}$$

where we have compactly abbreviated $\eta_{\mathbf{STS}^\dagger}(\gamma) = \eta$. *The corresponding Shannon transform is*

$$\mathcal{V}_{\mathbf{STS}^\dagger}(\gamma) = \beta\mathcal{V}_{\mathbf{T}}(\eta\gamma) + \log\frac{1}{\eta} + (\eta - 1)\log e\,. \tag{4.15}$$

Theorem 4.7 ([21]). *Define* $\mathbf{H} = \mathbf{CSA}$ *where* $\mathbf{S}$ *is an* $N \times K$ *matrix whose entries are i.i.d. complex random variables with variance* $\frac{1}{N}$. *Let* $\mathbf{C}$ *and* $\mathbf{A}$ *be, respectively,* $N \times N$ *and* $K \times K$ *random matrices such that the asymptotic spectra of* $\mathbf{D} = \mathbf{CC}^\dagger$ *and* $\mathbf{T} = \mathbf{AA}^\dagger$ *converge a.s. to a nonrandom limit. If* $\mathbf{C}$, $\mathbf{A}$ *and* $\mathbf{S}$ *are independent, as* $K, N \to \infty$ *with* $\frac{K}{N} \to \beta$, *the Shannon transform of* $\mathbf{HH}^\dagger$ *is given by:*

$$\mathcal{V}_{\mathbf{HH}^\dagger}(\gamma) = \mathcal{V}_{\mathbf{D}}(\beta\gamma_d) + \beta\mathcal{V}_{\mathbf{T}}(\gamma_t) - \beta\frac{\gamma_d\gamma_t}{\gamma}\log e \tag{4.16}$$

where

$$\frac{\gamma_d\gamma_t}{\gamma} = 1 - \eta_{\mathbf{T}}(\gamma_t) \qquad \beta\frac{\gamma_d\gamma_t}{\gamma} = 1 - \eta_{\mathbf{D}}(\beta\gamma_d) \tag{4.17}$$

while the η *transform of* $\mathbf{HH}^\dagger$ *can be obtained as*

$$\eta_{\mathbf{HH}^\dagger}(\gamma) = \eta_{\mathbf{D}}(\beta\,\gamma_d(\gamma)) \tag{4.18}$$

where $\gamma_d(\gamma)$ *is the solution to (4.17).*

The asymptotic fraction of zero eigenvalues of $\mathbf{HH}^\dagger$ *equals*

$$\lim_{\gamma\to\infty} \eta_{\mathbf{HH}^\dagger}(\gamma) = 1 - \min\{\beta\,\mathbb{P}[\mathsf{T} \neq 0], \mathbb{P}[\mathsf{D} \neq 0]\}\,.$$

Moreover, it has been proved in [3] and [21] that:

Theorem 4.8 ([21]). *Let* $\mathbf{H}$ *be an* $N \times K$ *matrix defined as in Theorem 4.7 whose* j*th column is* $\mathbf{h}_j$. *As* $K, N \to \infty$, *with* $\frac{K}{N} \to \beta$

$$\frac{1}{\|\mathbf{h}_j\|^2}\mathbf{h}_j^\dagger\left(\mathbf{I} + \gamma\sum_{\ell\neq j}\mathbf{h}_\ell\mathbf{h}_\ell^\dagger\right)^{-1}\mathbf{h}_j \xrightarrow{a.s.} \frac{\gamma_t(\gamma)}{\gamma\mathbb{E}[\mathsf{D}]} \tag{4.19}$$

with $\gamma_t(\gamma)$ *satisfying (4.17).*

According to (4.17), it is easy to verify that $\gamma_{\mathsf{t}}(\gamma)$ is the solution to

$$\gamma_{\mathsf{t}} = \mathbb{E}\left[\frac{\gamma \mathsf{D}}{1 + \gamma\,\beta\,\mathsf{D}\mathbb{E}\left[\frac{\mathsf{T}}{1+\mathsf{T}\,\gamma_{\mathsf{t}}}\right]}\right] \tag{4.20}$$

where for simplicity notation we have abbreviated $\gamma_{\mathsf{t}}(\gamma) = \gamma_{\mathsf{t}}$.

Notice that, given a linear memoryless vector channel as in (2.1) with the channel matrix $\mathbf{H}$ defined as in Theorem 4.7, the signal-to-interference-to-noise ratio SINR_k, incurred estimating the kth component of channel input based on its noisy received observations, is given by

$$\mathsf{SINR}_k = \mathsf{SNR}\,\mathbf{h}_j^\dagger \left(\mathbf{I} + \mathsf{SNR}\sum_{\ell \neq j} \mathbf{h}_\ell \mathbf{h}_\ell^\dagger\right)^{-1} \mathbf{h}_j \tag{4.21}$$

where SNR represents the transmitted signal-to-noise ratio.

Thus from Theorem 4.8, it follows that the multiuser efficiency of the kth user achieved by the MMSE receiver, $\eta_k^{\mathsf{MMSE}}(\mathsf{SNR})$, converges a.s. to:

$$\eta_k^{\mathsf{MMSE}}(\mathsf{SNR}) = \frac{\mathsf{SINR}_k}{\mathsf{SNR}\,\|\mathbf{h}_k\|^2} \tag{4.22}$$

$$\xrightarrow{a.s.} \frac{\gamma_{\mathsf{t}}(\mathsf{SNR})}{\mathsf{SNR}\,\mathbb{E}[\mathsf{D}]}. \tag{4.23}$$

A special case of Theorem 4.7 is when $\mathbf{H} = \mathbf{SA}$ (i.e $\mathbf{C} = \mathbf{I}$). Then according to (4.20) and (4.14), we have that

$$\gamma_{\mathsf{t}}(\gamma) = \gamma\,\eta_{\mathbf{STS}}(\gamma)$$

and consequently MMSE multiuser efficiency, $\eta_k^{\mathsf{MMSE}}(\mathsf{SNR})$ of a channel as in (2.1) with $\mathbf{H} = \mathbf{SA}$ converges a.s. to the η transform of $\mathbf{STS}$ evaluate at SNR:[b]

$$\eta_k^{\mathsf{MMSE}}(\mathsf{SNR}) \xrightarrow{a.s.} \eta_{\mathbf{STS}}(\mathsf{SNR}). \tag{4.24}$$

Theorem 4.9 ([21, 22]). *Let* $\mathbf{H}$ *be an* $N \times K$ *matrix defined as in Theorem 4.7. Defining*

$$\beta' = \beta \frac{\mathbb{P}[\mathsf{T} \neq 0]}{\mathbb{P}[\mathsf{D} \neq 0]},$$

$$\lim_{\gamma \to \infty}\left(\log(\gamma\,\beta) - \frac{\mathcal{V}_{\mathbf{HH}^\dagger}(\gamma)}{\min\{\beta\,\mathbb{P}[\mathsf{T} \neq 0], \mathbb{P}[\mathsf{D} \neq 0]\}}\right) = \mathcal{L}_\infty \tag{4.25}$$

[b]The conventional notation for multiuser efficiency is η (cf. [4]); the relationship in (5.6) is the motivation for the choice of the η transform terminology introduced in this section.

with

$$\mathcal{L}_\infty = \begin{cases} -\mathbb{E}\left[\log \frac{\mathbb{P}[\mathsf{T}\neq 0]\mathsf{D}'}{\alpha\beta' e}\right] - \beta'\mathcal{V}_{\mathsf{T}'}(\alpha) & \beta' > 1 \\ -\mathbb{E}\left[\log \frac{\mathsf{T}'\mathsf{D}'}{e}\right] & \beta' = 1 \\ -\mathbb{E}\left[\log \frac{\Gamma_\infty \mathsf{T}'}{e}\right] - \frac{1}{\beta'}\mathcal{V}_{\mathsf{D}'}\left(\frac{\mathbb{P}[\mathsf{T}\neq 0]}{\Gamma_\infty}\right) & \beta' < 1 \end{cases} \tag{4.26}$$

with α *and* Γ_∞*, respectively, solutions to*

$$\eta_{\mathsf{T}'}(\alpha) = 1 - \frac{1}{\beta'}, \qquad \eta_{\mathsf{D}'}\left(\frac{\mathbb{P}[\mathsf{T}\neq 0]}{\Gamma_\infty}\right) = 1 - \beta' \tag{4.27}$$

and with D' *and* T' *the restrictions of* D *and* T *to the events* $\mathsf{D} \neq 0$ *and* $\mathsf{T} \neq 0$.

The foregoing result gives the power offset (3.9) of the linear vector memoryless channel in (2.1) when $\mathbf{H}$ is defined as in Theorem 4.7.

Theorem 4.10. *As* $\gamma \to \infty$*, we have that*

$$\lim_{\gamma\to\infty} \frac{\gamma_t(\gamma)}{\gamma} = \beta\,\mathbb{P}[\mathsf{T} > 0]\,\Gamma_\infty \tag{4.28}$$

where $\gamma_t(\gamma)$ *is the solution to (4.17) while* Γ_∞ *is the solution to (4.27) for* $\beta' < 1$ *and* 0 *otherwise.*

Definition 4.11. An $N \times K$ matrix $\mathbf{P}$ is *asymptotically row-regular* if

$$\lim_{K\to\infty} \frac{1}{K}\sum_{j=1}^{K} 1\{\mathsf{P}_{i,j} \leq \alpha\}$$

is independent of i for all $\alpha \in \mathbb{R}$, as $K, N \to \infty$ and the aspect ratio $\frac{K}{N}$ converges to a constant. A matrix whose transpose is asymptotically row-regular is called *asymptotically column-regular*. A matrix that is both asymptotically row-regular and asymptotically column-regular is called *asymptotically doubly-regular* and satisfies

$$\lim_{N\to\infty} \frac{1}{N}\sum_{i=1}^{N} \mathsf{P}_{i,j} = \lim_{K\to\infty} \frac{1}{K}\sum_{j=1}^{K} \mathsf{P}_{i,j}. \tag{4.29}$$

If (4.29) is equal to 1, then $\mathbf{P}$ is *standard asymptotically doubly-regular*.

Theorem 4.12 ([21, 3]). *Define an $N \times K$ complex random matrix* $\mathbf{H}$ *whose entries are independent complex random variables (arbitrarily distributed) with identical means. Let their second moments be*

$$\mathbb{E}\left[|\mathsf{H}_{i,j}|^2\right] = \frac{\mathsf{P}_{i,j}}{N} \tag{4.30}$$

with $\mathbf{P}$ *an $N \times K$ deterministic standard asymptotically doubly-regular matrix whose entries are uniformly bounded for any N. The asymptotic empirical eigenvalue distribution of* $\mathbf{HH}^\dagger$ *converges a.s. to the Marčenko-Pastur distribution whose density is given by (4.8).*

Using Lemma 2.6 in [23], Theorem 4.12 can be extended to matrices whose mean has rank r where $r > 1$ but such that

$$\lim_{N\to\infty} \frac{r}{N} = 0\,.$$

Definition 4.13. Consider an $N \times K$ random matrix $\mathbf{H}$ whose entries have variances

$$\mathrm{Var}[\mathsf{H}_{i,j}] = \frac{\mathsf{P}_{i,j}}{N} \tag{4.31}$$

with $\mathbf{P}$ an $N \times K$ deterministic matrix whose entries are uniformly bounded. For each N, let

$$v^N : [0,1) \times [0,1) \to \mathbb{R}$$

be the *variance profile* function given by

$$v^N(x,y) = \mathsf{P}_{i,j} \qquad \frac{i-1}{N} \le x < \frac{i}{N}, \qquad \frac{j-1}{K} \le y < \frac{j}{K}\,. \tag{4.32}$$

Whenever $v^N(x,y)$ converges uniformly to a limiting bounded measurable function, $v(x,y)$, we define this limit as the *asymptotic variance profile* of $\mathbf{H}$.

Theorem 4.14 ([24–26]). *Let* $\mathbf{H}$ *be an $N \times K$ complex random matrix whose entries are independent zero-mean complex random variables (arbitrarily distributed) with variances*

$$\mathbb{E}\left[|\mathsf{H}_{i,j}|^2\right] = \frac{\mathsf{P}_{i,j}}{N} \tag{4.33}$$

where $\mathbf{P}$ *is an $N \times K$ deterministic matrix whose entries are uniformly bounded and from which the asymptotic variance profile of* $\mathbf{H}$*, denoted $v(x,y)$, can be obtained as per Definition 4.13. As $K, N \to \infty$ with $\frac{K}{N} \to \beta$,* *the empirical eigenvalue distribution of* $\mathbf{HH}^\dagger$ *converges a.s. to a limiting distribution whose* η *transform is*

$$\eta_{\mathbf{HH}^\dagger}(\gamma) = \mathbb{E}\left[\Gamma_{\mathbf{HH}^\dagger}(X,\gamma)\right] \tag{4.34}$$

with $\Gamma_{\mathbf{HH}^\dagger}(x,\gamma)$ *satisfying the equations,*

$$\Gamma_{\mathbf{HH}^\dagger}(x,\gamma) = \frac{1}{1+\beta\gamma\mathbb{E}[v(x,\mathsf{Y})\Upsilon_{\mathbf{HH}^\dagger}(\mathsf{Y},\gamma)]} \tag{4.35}$$

$$\Upsilon_{\mathbf{HH}^\dagger}(y,\gamma) = \frac{1}{1+\gamma\,\mathbb{E}[v(\mathsf{X},y)\Gamma_{\mathbf{HH}^\dagger}(\mathsf{X},\gamma)]} \tag{4.36}$$

where X *and* Y *are independent random variables uniform on* $[0,1]$.

The zero-mean hypothesis in Theorem 4.14 can be relaxed using Lemma 2.6 in [23]. Specifically, if the rank of $\mathbb{E}[\mathbf{H}]$ is $o(N)$, then Theorem 4.14 still holds.

Theorem 4.15 ([21]). *Let* $\mathbf{H}$ *be an* $N\times K$ *matrix defined as in Theorem 4.14. Further define*

$$\mathsf{F}^{(N)}(y,\gamma) = \frac{1}{\|\mathbf{h}_j\|^2}\mathbf{h}_j^\dagger\left(\mathbf{I}+\gamma\sum_{\ell\neq j}\mathbf{h}_\ell\mathbf{h}_\ell^\dagger\right)^{-1}\mathbf{h}_j, \qquad \frac{j-1}{K}\leq y<\frac{j}{K}.$$

As $K,N\to\infty$, $\mathsf{F}^{(N)}$ *converges a.s. to* $\frac{\mathsf{F}(y,\gamma)}{\mathbb{E}[v(\mathsf{X},y)]}$, *with* $\mathsf{F}(y,\gamma)$ *solution to the fixed-point equation*

$$\mathsf{F}(y,\gamma) = \mathbb{E}\left[\frac{v(\mathsf{X},y)}{1+\gamma\beta\mathbb{E}\left[\frac{v(\mathsf{X},\mathsf{Y})}{1+\gamma\mathsf{F}(\mathsf{Y},\gamma)}|\mathsf{X}\right]}\right] \qquad y\in[0,1]. \tag{4.37}$$

The Shannon transform of the asymptotic spectrum of $\mathbf{HH}^\dagger$ is given by the following result.

Theorem 4.16 ([27, 3]). *Let* $\mathbf{H}$ *be an* $N\times K$ *complex random matrix defined as in Theorem 4.14. The Shannon transform of the asymptotic spectrum of* $\mathbf{HH}^\dagger$ *is*

$$\begin{aligned}\mathcal{V}_{\mathbf{HH}^\dagger}(\gamma) &= \beta\,\mathbb{E}\left[\log(1+\gamma\,\mathbb{E}[v(\mathsf{X},\mathsf{Y})\Gamma_{\mathbf{HH}^\dagger}(\mathsf{X},\gamma)|\mathsf{Y}])\right]\\ &\quad+\mathbb{E}\left[\log(1+\gamma\,\beta\,\mathbb{E}[v(\mathsf{X},\mathsf{Y})\Upsilon_{\mathbf{HH}^\dagger}(\mathsf{Y},\gamma)|\mathsf{X}])\right]\\ &\quad-\gamma\,\beta\,\mathbb{E}\left[v(\mathsf{X},\mathsf{Y})\Gamma_{\mathbf{HH}^\dagger}(\mathsf{X},\gamma)\Upsilon_{\mathbf{HH}^\dagger}(\mathsf{Y},\gamma)\right]\log e\end{aligned} \tag{4.38}$$

with $\Gamma_{\mathbf{HH}^\dagger}(\cdot,\cdot)$ *and* $\Upsilon_{\mathbf{HH}^\dagger}(\cdot,\cdot)$ *satisfying (4.35) and (4.36).*

Theorem 4.17 ([21]). *Let $\mathbf{H}$ be an $N \times K$ complex random matrix defined as in Theorem 4.14. Then, denoting*

$$\beta' = \beta \frac{\mathbb{P}[\mathbb{E}[v(\mathsf{X},\mathsf{Y})|\mathsf{Y}] \neq 0]}{\mathbb{P}[\mathbb{E}[v(\mathsf{X},\mathsf{Y})|\mathsf{X}] \neq 0]},$$

we have that

$$\lim_{\gamma\to\infty}\left(\log(\gamma\beta) - \frac{\mathcal{V}_{\mathbf{HH}^\dagger}(\gamma)}{\min\{\beta\mathbb{P}[\mathbb{E}[v(\mathsf{X},\mathsf{Y})|\mathsf{Y}]\neq 0], \mathbb{P}[\mathbb{E}[v(\mathsf{X},\mathsf{Y})|\mathsf{X}]\neq 0]\}}\right) = \mathcal{L}_\infty$$

with

$$\mathcal{L}_\infty \overset{a.s.}{\to} \begin{cases} -\mathbb{E}\left[\log\left(\frac{1}{e}\mathbb{E}\left[\frac{v(\mathsf{X}',\mathsf{Y}')}{1+\alpha(\mathsf{Y}')}|\mathsf{X}'\right]\right)\right] - \beta'\mathbb{E}\left[\log\left(1+\alpha(\mathsf{Y}')\right)\right] & \beta' > 1 \\ -\mathbb{E}\left[\log\frac{v(\mathsf{X}',\mathsf{Y}')}{e}\right] & \beta' = 1 \\ -\mathbb{E}\left[\log\frac{\Gamma_\infty(\mathsf{Y}')}{e}\right] - \frac{1}{\beta'}\mathbb{E}\left[\log\left(1+\mathbb{E}\left[\frac{v(\mathsf{X}',\mathsf{Y}')}{\Gamma_\infty(\mathsf{Y}')}|\mathsf{X}'\right]\right)\right] & \beta' < 1 \end{cases}$$

with X' and Y' the restrictions of X and Y to the events $\mathbb{E}[v(\mathsf{X},\mathsf{Y})|\mathsf{X}]\neq 0$ and $\mathbb{E}[v(\mathsf{X},\mathsf{Y})|\mathsf{Y}]\neq 0$, respectively. The function $\alpha(\cdot)$ is the solution, for $\beta'>1$, of

$$\alpha(y) = \frac{1}{\beta'}\mathbb{E}\left[\frac{v(\mathsf{x}',y)}{\mathbb{E}\left[\frac{v(\mathsf{R}',\mathsf{Y}')}{1+\alpha(\mathsf{Y}')}|\mathsf{x}'\right]}\right] \tag{4.39}$$

whereas $\Gamma_\infty(\cdot)$ is the solution, for $\beta'<1$, of

$$\mathbb{E}\left[\frac{1}{1+\mathbb{E}\left[\frac{v(\mathsf{X}',\mathsf{Y}')}{\Gamma_\infty(\mathsf{Y}')}|\mathsf{X}'\right]}\right] = 1-\beta'. \tag{4.40}$$

As we will see in the next section, Theorems 4.14–4.17 give the MMSE performance, the mutual information and the power offset of a large class of vector channel of interest in wireless communications which are described by random matrices with either correlated or independent entries.

Let **STS** be an $N \times N$ random matrix with $\mathbf{S}$ and $\mathbf{T}$ be respectively $N \times K$ and $K \times K$ random matrices as stated in Theorem 4.6. We have seen that the ESD of **STS** converges a.s. to a nonrandom limit whose Shannon and η transform satisfy (4.15) and (4.14) respectively.

From this limiting behavior of the ESD of **STS**, it follows immediately that linear spectral statistics of the form:

$$\frac{1}{N}\sum_{i=1}^{N} g(\lambda_i) \tag{4.41}$$

with $g(\cdot)$ a continuous function on the real line with bounded and continuous derivatives, converge a.s. to a nonrandom quantity. A recent central result in random matrix theory by Bai and Silverstein (2004) [17] shows their rate of convergence to be $1/N$. Moreover, they show that:

Theorem 4.18 ([17]). *Let* $\mathbf{S}$ *be an* $N \times K$ *complex matrix defined as in Theorem 4.6 and such that its* (i, j)*th entry satisfies:*

$$\mathbb{E}[\mathsf{S}_{i,j}] = 0 \qquad \mathbb{E}[|\mathsf{S}_{i,j}|^4] = \frac{2}{N^2}\,. \tag{4.42}$$

Let $\mathbf{T}$ *be a* $K \times K$ *matrix defined as in Theorem 4.6 whose spectral norm is bounded. Let* $g(\cdot)$ *be a continuous function on the real line with bounded and continuous derivatives, analytic on a open set containing the interval*[c]

$$[\liminf_K \phi_K \max{}^2\{0, 1-\sqrt{\beta}\}, \limsup_K \phi_1(1+\sqrt{\beta})^2]$$

where $\phi_1 \geq \cdots \geq \phi_K$ *are the eigenvalues of* $\mathbf{T}$. *Denoting by* λ_i *and* $\mathsf{F}_{\mathbf{STS}^\dagger}(\cdot)$, *respectively, the* i*th eigenvalue and the asymptotic ESD of* $\mathbf{STS}^\dagger$, *the random variable*

$$\mathbf{\Delta}_N = \sum_{i=1}^{N} g(\lambda_i) - N\int g(x)\, d\mathsf{F}_{\mathbf{STS}^\dagger} \tag{4.43}$$

converges, as $K, N \to \infty$ *with* $\frac{K}{N} \to \beta$, *to a zero-mean Gaussian random variable with variance*

$$\mathbb{E}[\mathbf{\Delta}^2] = -\frac{1}{2\pi^2}\oint\oint \frac{\dot{g}(\mathcal{Z}(\sigma_1))g(\mathcal{Z}(\sigma_2))}{\sigma_2-\sigma_1} d\sigma_1 d\sigma_2 \tag{4.44}$$

where $\dot{g}(x) = \frac{d}{dx}g(x)$ *while*

$$\mathcal{Z}(\sigma) = -\frac{1}{\sigma}\left(1-\beta(1-\eta_{\mathbf{T}}(\sigma))\right)\,. \tag{4.45}$$

In (5.64) the integration variables σ_1 *and* σ_2 *follow closed contours, which we may take to be non-overlapping and counterclockwise, such that the corresponding contours mapped through* $\mathcal{Z}(\sigma)$ *enclose the support of* $\mathsf{F}_{\mathbf{STS}^\dagger}(\cdot)$.

[c] In [28] this interval contains the spectral support of $\mathbf{S}^\dagger\mathbf{ST}$.

Using the foregoing result, we have that

Theorem 4.19 ([29]). *Let* $\mathbf{S}$ *be an* $N \times K$ *complex matrix defined as in Theorem 4.20. Let* $\mathbf{T}$ *be an Hermitian random matrix independent of* $\mathbf{S}$ *with bounded spectral norm and whose asymptotic ESD converges a.s. to a nonrandom limit. Denote by* $\mathcal{V}_{\mathbf{STS}^\dagger}(\gamma)$ *the Shannon transform of* $\mathbf{STS}^\dagger$*. As* $K, N \to \infty$ *with* $\frac{K}{N} \to \beta$*, the random variable*

$$\Delta_N = \log\det(\mathbf{I} + \gamma\mathbf{STS}^\dagger) - N\mathcal{V}_{\mathbf{STS}^\dagger}(\gamma) \tag{4.46}$$

is asymptotically zero-mean Gaussian with variance

$$\mathbb{E}[\Delta^2] = -\log\left(1 - \beta\,\mathbb{E}\left[\left(\frac{\mathsf{T}\gamma\eta_{\mathbf{STS}^\dagger}(\gamma)}{1+\mathsf{T}\gamma\eta_{\mathbf{STS}^\dagger}(\gamma)}\right)^2\right]\right)$$

where the expectation is over the nonnegative random variable T *whose distribution is given by the asymptotic ESD of* $\mathbf{T}$*.*

From Jensen's inequality and (4.14), a tight lower bound for the variance of Δ in Theorem 4.19 is given by [3, Eq. 2.239]:

$$\mathbb{E}[\Delta^2] \geq -\log\left(1 - \frac{(1-\eta_{\mathbf{STS}^\dagger}(\gamma))^2}{\beta}\right) \tag{4.47}$$

with strict equality if $\mathbf{T} = \mathbf{I}$. In fact, Theorem 4.19 can be particularized to the case $\mathbf{T} = \mathbf{I}$ to obtain:

Theorem 4.20. *Let* $\mathbf{S}$ *be an* $N \times K$ *complex as in Theorem 4.19. As* $K, N \to \infty$ *with* $\frac{K}{N} \to \beta$*, the random variable*

$$\Delta_N = \log\det(\mathbf{I} + \gamma\mathbf{SS}^\dagger) - N\mathcal{V}_{\mathbf{SS}^\dagger}(\gamma) \tag{4.48}$$

is asymptotically zero-mean Gaussian with variance

$$\mathbb{E}[\Delta^2] = -\log\left(1 - \frac{(1-\eta_{\mathbf{SS}^\dagger}(\gamma))^2}{\beta}\right) \tag{4.49}$$

where $\eta_{\mathbf{SS}^\dagger}(\gamma)$ *and* $\mathcal{V}_{\mathbf{HH}^\dagger}(\gamma)$ *are given in (4.9) and (4.10).*

5. Applications to Wireless Communications

In this section we focus our attention on some of the major wireless channels that are simple yet practically very relevant and able to capture various features of contemporary interest:

A. Randomly spread Code Division Multiple Access (CDMA) channels subject to either frequency-flat or frequency-selective fading.

B. Single-user multiantenna channels subject to frequency-flat fading.

Naturally, random matrices also arise in models that incorporate more than one of the above features (multiuser, multiantenna, fading, wideband). Although realistic models do include several (if not all) of the above features it is conceptually advantageous to start by deconstructing them into their essential ingredients.

In the next subsections we describe the foregoing scenarios and show how the distribution of the squared singular values of certain matrices determine communication limits in both the coded regime (Shannon capacity) and the uncoded regime (probability of error). Each of above channels are analyzed in the asymptotic regime where K (number of transmit antennas or number of users) and N (number of receive antennas or number of degrees of freedom per symbol in the CDMA channel) go to infinity while the ratio goes to a constant. In such regime, for each of these channels, we derive several performance measures of engineering interest which are determined by the distribution of the singular values of the channel matrix.

Unless otherwise stated, the analysis applies to coherent reception and thus it is presumed that the state of the channel is perfectly tracked by the receiver. The degree of channel knowledge at the transmitter, on the other hand, as well as the rapidity of the fading fluctuations (ergodic or non-ergodic regime) are specified for each individual setting.

5.1. *CDMA*

An application that is very suitable is the code-division multiple access channel or CDMA channel, were each user is assigned a signature vector known at the receiver which can be seen as an element of an N dimensional signal space. Based on the nature of this signal space we can distinguish between:

- Direct sequence CDMA used in many current cellular systems (IS-95, cdma2000, UMTS)
- Multi-carrier CDMA being considered for fourth generation of cellular systems.

5.1.1. *DS-CDMA frequency-flat fading*

Concerning the DS-CDMA, we first focus on channels whose response is flat over the signal bandwidth which implies that the received signature of each user is just a scaled version of the transmitted one where the scaling factors are the independent fading coefficients for each user.

Considering the basic synchronous DS-CDMA [4] with K users and spreading factor N in a frequency-flat fading environment, the vector $\mathbf{x}$ contains the symbols transmitted by the K users while the role of $\mathbf{H}$ is played by the product of two matrices, $\mathbf{S}$ and $\mathbf{A}$, where $\mathbf{S}$ is a $N \times K$ matrix whose columns are the spreading sequences

$$\mathbf{S} = [\,\mathbf{s}_1 \,|\, \dots \,|\mathbf{s}_K\,] \tag{5.1}$$

and $\mathbf{A}$ is a $K \times K$ diagonal matrix whose kth diagonal entry is the complex fading coefficient of kth user. The model thus specializes to

$$\mathbf{y} = \mathbf{SAx} + \mathbf{n}. \tag{5.2}$$

The standard random signature model [4] assumes that the entries of $\mathbf{S}$, are chosen independently and equiprobably on $\{-\frac{1}{\sqrt{N}}, \frac{1}{\sqrt{N}}\}$. Moreover, the random signature model is often generalized to encompass non-binary (e.g. Gaussian) distributions for the amplitudes that modulate the chip waveforms. With that, the randomness in the received sequence can also reflect the impact of fading. One motivation for modeling the signatures random is the use of "long sequences" in some commercial CDMA systems, where the period of the pseudo-random sequence spans many symbols. Another motivation is to provide a baseline of comparison for systems that use signature waveform families with low cross-correlations.

The arithmetic mean of the MMSE's for the K users satisfies [4]

$$\frac{1}{K}\sum_{k=1}^{K} \mathsf{MMSE}_k = \frac{1}{K}\operatorname{tr}\left\{(\mathbf{I} + \mathsf{SNR}\,\mathbf{A}^\dagger\mathbf{S}^\dagger\mathbf{SA})^{-1}\right\} \tag{5.3}$$

$$\to \eta_{\mathbf{A}^\dagger\mathbf{S}^\dagger\mathbf{SA}}(\mathsf{SNR}) \tag{5.4}$$

whereas the MMSE multiuser efficiency of the kth user, $\eta_k^{\mathsf{MMSE}}(\mathsf{SNR})$, given in (3.17) is:

$$\eta_k^{\mathsf{MMSE}}(\mathsf{SNR}) = \mathbf{s}_k^T\left(\mathbf{I} + \sum_{i\neq k}\mathsf{SNR}\,|\mathsf{A}_i|^2\mathbf{s}_i\mathbf{s}_i^T\right)^{-1}\mathbf{s}_k \tag{5.5}$$

$$\to \eta_{\mathbf{SAA}^\dagger\mathbf{S}^\dagger}(\mathsf{SNR}) \tag{5.6}$$

where the limit follows from (4.24). According to Theorem 4.6, the MMSE multiuser efficiency, abbreviated as

$$\eta = \eta_{\mathbf{SAA}^\dagger\mathbf{S}^\dagger}(\mathsf{SNR}), \tag{5.7}$$

is the solution to the fixed-point equation

$$1 - \eta = \beta \left(1 - \eta_{|\mathsf{A}|^2}(\mathsf{SNR}\,\eta)\right), \tag{5.8}$$

where $\eta_{|\mathsf{A}|^2}$ is the η transform of the asymptotic empirical distribution of $\{|\mathsf{A}_1|^2, \ldots, |\mathsf{A}_K|^2\}$.

The capacity of the optimum receiver (normalized by the spreading factor N) which is given by [4]:

$$\mathcal{C}^{\mathsf{opt}}(\beta\,\mathsf{SNR}) = \lim_{N\to\infty} \frac{1}{N} \log\det\left(\mathbf{I} + \mathsf{SNR}\,\mathbf{SAA}^\dagger\mathbf{S}^\dagger\right),$$

according to Theorem 4.6, equals (4.15).

It has been proved in [30] that in the asymptotic regime the normalized spectral efficiency of the MMSE receiver converges to

$$\mathcal{C}^{\mathsf{MMSE}}(\beta\,\mathsf{SNR}) = \lim_{N\to\infty} \frac{1}{N} \sum_{k=1}^{K} \mathbb{E}\left[\log\left(1 + \mathsf{SINR}_k\right)\right] \tag{5.9}$$

from which it follows using (4.22) that

$$\mathcal{C}^{\mathsf{MMSE}}(\beta\,\mathsf{SNR}) = \beta\,\mathbb{E}\left[\log\left(1 + |\mathsf{A}|^2\,\mathsf{SNR}\,\eta_{\mathbf{SAA}^\dagger\mathbf{S}^\dagger}(\mathsf{SNR})\right)\right]. \tag{5.10}$$

Based on (5.10), the capacity of the optimum receiver can be characterized in terms of the MMSE spectral efficiency [30]:

$$\begin{aligned}\mathcal{C}^{\mathsf{opt}}(\beta\,\mathsf{SNR}) = \mathcal{C}^{\mathsf{MMSE}}(\beta\,\mathsf{SNR}) + \log \frac{1}{\eta_{\mathbf{SAA}^\dagger\mathbf{S}^\dagger}(\mathsf{SNR})} \\ + \left(\eta_{\mathbf{SAA}^\dagger\mathbf{S}^\dagger}(\mathsf{SNR}) - 1\right)\log e.\end{aligned} \tag{5.11}$$

The unfaded equal power case is obtained by the the above model assuming $\mathbf{A} = A\mathbf{I}$, where A is the transmitted amplitude equal for all users. In this case, the channel matrix in (5.24) has independent identically distributed entries and thus, according to Theorem 4.3, its asymptotic ESD converges to the Marčenko-Pastur law. Thus the normalized capacity achieved with the optimum receiver in the asymptotic regime is (cf. Theorem 4.4):

$$\begin{aligned}\mathcal{C}^{\mathsf{opt}}(\beta, \mathsf{SNR}) = \beta \log\left(1 + \mathsf{SNR} - \frac{\mathcal{F}(\mathsf{SNR}, \beta)}{4}\right) \\ + \log\left(1 + \mathsf{SNR}\,\beta - \frac{\mathcal{F}(\mathsf{SNR}, \beta)}{4}\right) - \frac{\mathcal{F}(\mathsf{SNR}, \beta)}{4\,\mathsf{SNR}} \log e,\end{aligned} \tag{5.12}$$

while the MMSE converges to

$$1 - \frac{\mathcal{F}(\mathsf{SNR}, \beta)}{4\ \mathsf{SNR}\,\beta} \tag{5.13}$$

with $\mathcal{F}(\cdot,\cdot)$ defined in (4.11). Using (4.24) and (4.9), the maximum SINR (achieved by the MMSE linear receiver) converges to [4]

$$\mathsf{SNR} - \frac{\mathcal{F}(\mathsf{SNR}, \beta)}{4}. \tag{5.14}$$

Let us consider a synchronous DS-CDMA downlink with K active users employing random spreading codes and operating over a frequency-selective fading channel. Then $\mathbf{H}$ in (2.1) particularizes to

$$\mathbf{H} = \mathbf{CSA} \tag{5.15}$$

where $\mathbf{A}$ is a $K \times K$ deterministic diagonal matrix containing the amplitudes of the users and $\mathbf{C}$ is an $N \times N$ Toeplitz matrix defined as

$$(\mathbf{C})_{i,j} = \frac{1}{W_c} c\left(\frac{i-j}{W_c}\right) \tag{5.16}$$

with $c(\cdot)$ the impulse response of the channel.

Using Theorem 4.8 and with the aid of an auxiliary function $\chi(\mathsf{SNR})$, abbreviated as χ, we obtain that the MMSE multiuser efficiency of the kth user, abbreviated as $\eta = \eta^{\mathsf{MMSE}}(\mathsf{SNR})$, is the solution to

$$\beta\,\eta\,\chi = \frac{1 - \eta_{|\mathsf{C}|^2}(\beta\,\chi)}{\mathbb{E}[|\mathsf{C}|^2]} \tag{5.17}$$

$$\eta\,\chi = \frac{1 - \eta_{|\mathsf{A}|^2}(\mathsf{SNR}\,\mathbb{E}[|\mathsf{C}|^2]\eta)}{\mathbb{E}[|\mathsf{C}|^2]} \tag{5.18}$$

where $|\mathsf{C}|^2$ and $|\mathsf{A}|^2$ are independent random variables with distributions given by the asymptotic spectra of $\mathbf{CC}^\dagger$ and $\mathbf{AA}^\dagger$, respectively, while $\eta_{|\mathsf{C}|^2}(\cdot)$ and $\eta_{|\mathsf{A}|^2}(\cdot)$ represent their respective η transforms. Note that, using (4.20), instead of (5.18) and (5.17), we may write [31, 32]

$$\eta = \mathbb{E}\left[\frac{|\mathsf{C}|^2}{1 + \beta\ \mathsf{SNR}\,|\mathsf{C}|^2 \mathbb{E}\left[\frac{|\mathsf{A}|^2}{1 + \mathsf{SNR}\,|\mathsf{A}|^2\eta}\right]}\right]. \tag{5.19}$$

From Theorem 4.7 and (4.3) we have that:

$$\begin{aligned}\frac{1}{K}\sum_{k=1}^{K}\mathsf{MMSE}_k &= \frac{1}{K}\operatorname{tr}\left\{\left(\mathbf{I}+\mathsf{SNR}\,\mathbf{H}^\dagger\mathbf{H}\right)^{-1}\right\}\\ &= 1-\frac{1}{\beta}+\frac{1}{\beta}\eta_{|\mathsf{C}|^2}\left(\beta\chi(\mathsf{SNR})\right) \qquad (5.20)\end{aligned}$$

with $\chi(\cdot)$ solution to (5.18) and (5.17).

The special case of (5.19) for equal-power users was given in [33].

For contributions on the asymptotic analysis of the uplink DS-CDMA systems in frequency selective fading channels see [3,27,32]. In the context of CDMA channels the asymptotic random matrix theory find also application in channel estimation and design of reduced-complexity receivers (see [3] for tutorial overview of this topic).

5.1.2. *Multi-carrier CDMA*

If the channel is not flat over the signal bandwidth, then the received signature of each user is not simply a scaled version of the transmitted one.

In this case, we can insert suitable transmit and receive interfaces and choose the signature space in such a way that the equivalent channel that encompasses the actual channel plus the interfaces can be modeled as a random matrix $\mathbf{H}$ given by:

$$\begin{aligned}\mathbf{H} &= \mathbf{C}\circ\mathbf{SA} \qquad (5.21)\\ &= \mathbf{G}\circ\mathbf{S} \qquad (5.22)\end{aligned}$$

where $\circ$ denotes the Hadamard (element-wise) product [34], $\mathbf{S}$ is the random signature matrix in the frequency domain, while $\mathbf{G}$ is an $N \times K$ matrix whose columns are independent N-dimensional random vectors whose (ℓ, k)th element is given by

$$\mathbf{G}_{i,j} = |\mathsf{C}_{i,j}|^2\,|\mathsf{A}_j|^2 \qquad (5.23)$$

where A_k indicates the received amplitude of that kth user, which accounts for its average path loss, and $\mathsf{C}_{\ell,k}$ denotes the fading for the ℓth subcarrier of the kth user, independent across the users. For this scenario, the linear model (2.1) specializes to

$$\mathbf{y} = (\mathbf{G}\circ\mathbf{S})\mathbf{x}+\mathbf{n}. \qquad (5.24)$$

The SINR at the output of the MMSE receiver is

$$\mathsf{SINR}_k^{\mathsf{MMSE}} = \mathsf{SNR}\,|\mathsf{A}_k|^2 (\mathbf{c}_k \circ \mathbf{s}_k)^\dagger \left(\mathbf{I} + \mathsf{SNR}\,\mathbf{H}_k \mathbf{H}_k^\dagger\right)^{-1} (\mathbf{c}_k \circ \mathbf{s}_k)$$

where $\mathbf{H}_k$ indicates the matrix $\mathbf{H}$ with the kth column removed.

Let $v(\cdot,\cdot)$ be the two-dimensional channel profile of $\mathbf{G}$. Using Theorems 4.15 and (4.22), the multiuser efficiency is given by the following result.

Theorem 5.1 ([27]). *For $0 \le y \le 1$, the multiuser efficiency of the MMSE receiver for the $\lfloor yK \rfloor$th user converges a.s., as $K, N \to \infty$ with $\frac{K}{N} \to \beta$, to*

$$\lim_{K\to\infty} \eta^{\mathsf{MMSE}}_{\lfloor yK \rfloor}(\mathsf{SNR}) = \frac{\Psi(y, \mathsf{SNR})}{\mathbb{E}\left[v(\mathsf{X}, y)\right]} \tag{5.25}$$

where $\Psi(\cdot,\cdot)$ is a positive function solution to

$$\Psi(y, \mathsf{SNR}) = \mathbb{E}\left[\frac{v(\mathsf{X}, y)}{1 + \mathsf{SNR}\ \beta\mathbb{E}\left[\frac{v(\mathsf{X}, \mathsf{Y})}{1 + \mathsf{SNR}\,\Psi(\mathsf{Y}, \mathsf{SNR})}\Big|\mathsf{X}\right]}\right] \tag{5.26}$$

and the expectations are with respect to independent random variables X and Y both uniform on [0,1].

Most quantities of interest such as the multiuser efficiency and the capacity approach their asymptotic behaviors very rapidly as K and N grow large. Hence, we can get an extremely accurate approximation of the multiuser efficiency and consequently of the capacity with an arbitrary number of users, K, and a finite processing gain, N, simply by resorting to their asymptotic approximation with $v(x, y)$ replaced in Theorem 5.1 by

$$v(x, y) \approx |\mathsf{A}_k|^2\, |\mathsf{C}_{\ell,k}|^2 \qquad \frac{\ell-1}{N} \le x < \frac{\ell}{N} \quad \frac{k-1}{K} \le y < \frac{k}{K}.$$

Thus, we have that the multiuser efficiency of uplink MC-CDMA is closely approximated by

$$\eta_k^{\mathsf{MMSE}}(\mathsf{SNR}) \approx \frac{\Phi_k^N(\mathsf{SNR})}{\frac{1}{N}\sum_{\ell=1}^{N} |\mathsf{C}_{\ell,k}|^2} \tag{5.27}$$

with

$$\Phi_k^N(\mathsf{SNR}) = \frac{1}{N}\sum_{\ell=1}^{N} \frac{|\mathsf{C}_{\ell,k}|^2}{1 + \mathsf{SNR}\,\frac{\beta}{K}\sum_{j=1}^{K}\frac{|\mathsf{A}_j|^2}{1 + \mathsf{SNR}\,\Phi_j^N(\mathsf{SNR})}}. \tag{5.28}$$

From (5.9) using Theorem 5.1, the MMSE spectral efficiency converges, as $K, N \to \infty$, to

$$\mathcal{C}^{\mathsf{MMSE}}(\beta, \mathsf{SNR}) = \beta \mathbb{E}\left[\log\left(1 + \mathsf{SNR}\, \Psi(\mathsf{Y}, \mathsf{SNR})\right)\right] \tag{5.29}$$

where the function $\Psi(\cdot, \cdot)$ is the solution of (5.26).

As an application of Theorem 4.16, the capacity of a multicarrier CDMA channel is obtained.

Theorem 5.2 ([27]). *The capacity of the optimum receiver is*

$$\begin{aligned}\mathcal{C}^{\mathsf{opt}}(\beta, \mathsf{SNR}) = {} & \mathcal{C}^{\mathsf{MMSE}}(\beta, \mathsf{SNR}) \\ & + \mathbb{E}\left[\log(1 + \mathsf{SNR}\, \beta \mathbb{E}\left[v(\mathsf{X}, \mathsf{Y})\Upsilon(\mathsf{Y}, \mathsf{SNR}) | \mathsf{X}\right]\right] \\ & - \beta\, \mathsf{SNR}\; \mathbb{E}\left[\Psi(\mathsf{Y}, \mathsf{SNR})\Upsilon(\mathsf{Y}, \mathsf{SNR})\right] \log e \end{aligned} \tag{5.30}$$

with $\Psi(\cdot, \cdot)$ and $\Upsilon(\cdot, \cdot)$ satisfying the coupled fixed-point equations

$$\Psi(y, \mathsf{SNR}) = \mathbb{E}\left[\frac{v(\mathsf{X}, y)}{1 + \beta\, \mathsf{SNR}\, \mathbb{E}[v(\mathsf{X}, \mathsf{Y})\Upsilon(\mathsf{Y}, \mathsf{SNR}) | \mathsf{X}]}\right] \tag{5.31}$$

$$\Upsilon(y, \mathsf{SNR}) = \frac{1}{1 + \mathsf{SNR}\; \Psi(y, \mathsf{SNR})} \tag{5.32}$$

where X and Y are independent random variables uniform on $[0, 1]$.

Note that (5.30) appears as function of quantities with immediate engineering meaning. More precisely, $\mathsf{SNR}\, \Psi(y, \mathsf{SNR})$ is easily recognized from Theorem 5.1 as the SINR exhibited by the $\lfloor yK \rfloor$th user at the output of a linear MMSE receiver. In turn $\Upsilon(y, \mathsf{SNR})$ is the corresponding mean-square error. An alternative characterization of the capacity (inspired by the optimality by successive cancellation with MMSE protection against uncancelled users) is given by

$$\mathcal{C}^{\mathsf{opt}}(\beta, \mathsf{SNR}) = \beta \mathbb{E}\left[\log(1 + \mathsf{SNR}\, \daleth(\mathsf{Y}, \mathsf{SNR}))\right] \tag{5.33}$$

where

$$\daleth(y, \mathsf{SNR}) = \mathbb{E}\left[\frac{v(\mathsf{X}, y)}{1 + \mathsf{SNR}\, \beta(1 - y)\mathbb{E}\left[\frac{v(\mathsf{X}, \mathsf{Z})}{1 + \mathsf{SNR}\, \daleth(\mathsf{Z}, \mathsf{SNR})} | \mathsf{X}\right]}\right] \tag{5.34}$$

where X, and Z are independent random variables uniform on $[0, 1]$ and $[y, 1]$, respectively.

For the downlink, the structure of the transmitted MC-CDMA signal is identical to that of the uplink, but the difference with (5.21) is that every user experiences the same channel and thus $\mathbf{c}_k = \mathbf{c}$ for all $1 \le k \le K$. As a result,

$$\mathbf{H} = \mathbf{CSA}$$

with $\mathbf{C} = \mathrm{diag}(\mathsf{c})$ and $\mathbf{A} = \mathrm{diag}(\mathsf{a})$. Consequently Theorems 4.7–4.10 can be used for the asymptotic analysis of MC-CDMA downlink. For an extended survey on contributions on the asymptotic analysis of MC-CDMA channels see [3] and references therein.

5.2. *Multi-antenna channels*

Let us now consider a single-user channel where the transmitter has n_T antennas and the receiver has n_R antennas.

In this case, $\mathbf{x}$ contains the symbols transmitted from the n_T transmit antennas and $\mathbf{y}$ the symbols received by the n_R receive antennas. With frequency-flat fading, the entries of $\mathbf{H}$ represent the fading coefficients between each transmit and each receive antenna, typically modelled as zero-mean complex Gaussian and normalized such that

$$\mathbb{E}\left[\mathrm{tr}\{\mathbf{H}\mathbf{H}^\dagger\}\right] = n_\mathrm{R}\,. \tag{5.35}$$

If all antennas are co-polarized, the entries of $\mathbf{H}$ are identically distributed and thus the resulting variance of each entry is $\frac{1}{n_\mathrm{T}}$. (See [16, 35] for the initial contributions on this topic and [36–40] for recent articles of tutorial nature.)

In contrast with the multiaccess scenarios, in this case the signals transmitted by different antennas can be advantageously correlated and thus the covariance of $\mathbf{x}$ becomes relevant. Normalized by its energy per dimension, the input covariance is denoted by

$$\mathbf{\Phi} = \frac{\mathbb{E}[\mathbf{x}\mathbf{x}^\dagger]}{\frac{1}{n_\mathrm{T}}\mathbb{E}[\|\mathbf{x}\|^2]} \tag{5.36}$$

where the normalization ensures that $\mathbb{E}[\mathrm{tr}\{\mathbf{\Phi}\}] = n_\mathrm{T}$. It is useful to decompose this input covariance in its eigenvectors and eigenvalues, $\mathbf{\Phi} = \mathbf{V}\mathbf{P}\mathbf{V}^\dagger$. Each eigenvalue represents the (normalized) power allocated to the corresponding signalling eigenvector. Associated with $\mathbf{P}$, we define an input *power profile*

$$\mathcal{P}^{(n_\mathrm{R})}(t, \mathsf{SNR}) = \mathsf{P}_{j,j} \qquad \frac{j}{n_\mathrm{R}} \le t < \frac{j+1}{n_\mathrm{R}}$$

supported on $t \in (0, \beta]$. This profile specifies the power allocation at each SNR. As the number of antennas is driven to infinity, $\mathcal{P}^{(n_R)}(t, \mathsf{SNR})$ converges uniformly to a nonrandom function, $\mathcal{P}(t, \mathsf{SNR})$, which we term *asymptotic power profile.*

The capacity per receive antenna is given by the maximum over $\mathbf{\Phi}$ of the Shannon transform of the averaged empirical distribution of $\mathbf{H\Phi H}^\dagger$, i.e.

$$\mathcal{C}(\mathsf{SNR}) = \max_{\mathbf{\Phi}:\mathrm{tr}\,\mathbf{\Phi}=n_T} \mathcal{V}_{\mathbf{H\Phi H}^\dagger}(\mathsf{SNR}) \tag{5.37}$$

where

$$\mathsf{SNR} = \frac{\mathbb{E}[\|\mathbf{x}\|^2]}{\frac{1}{n_R}\mathbb{E}[\|\mathbf{n}\|^2]}. \tag{5.38}$$

If full CSI is available at the transmitter, then $\mathbf{V}$ should coincide with the eigenvector matrix of $\mathbf{H}^\dagger\mathbf{H}$ and $\mathbf{P}$ should be obtained through a waterfill process on the eigenvalues of $\mathbf{H}^\dagger\mathbf{H}$ [16, 41–43]. The resulting jth diagonal entry of $\mathbf{P}$ is

$$\mathsf{P}_{j,j} = \left(\nu - \frac{1}{\mathsf{SNR}\,\lambda_j(\mathbf{H}^\dagger\mathbf{H})}\right)^+ \tag{5.39}$$

where ν is such that $\mathrm{tr}\{\mathbf{P}\} = n_T$. Then, substituting in (5.37),

$$\mathcal{C}(\mathsf{SNR}) = \frac{1}{n_R}\log\det(\mathbf{I} + \mathsf{SNR}\,\mathbf{P\Lambda}) \tag{5.40}$$

$$= \beta\int(\log(\mathsf{SNR}\,\nu\lambda))^+ d\mathsf{F}^{n_T}_{\mathbf{H}^\dagger\mathbf{H}}(\lambda) \tag{5.41}$$

with $\mathbf{\Lambda}$ equal to the diagonal eigenvalue matrix of $\mathbf{H}^\dagger\mathbf{H}$.

If, instead, only statistical CSI is available, then $\mathbf{V}$ should be set, for all the channels that we will consider, to coincide with the eigenvectors of $\mathbb{E}[\mathbf{H}^\dagger\mathbf{H}]$ while the capacity-achieving power allocation, $\mathbf{P}$, can be found iteratively [44].

5.3. *Separable correlation model*

Antenna correlation at the transmitter and at the receiver, that is, between the columns and between the rows of $\mathbf{H}$, respectively, can be accounted for through corresponding correlation matrices $\mathbf{\Theta}_T$ and $\mathbf{\Theta}_R$ [45–47]. According to this model, which is referred to as separable correlation model, an $n_R \times n_T$ matrix $\mathbf{H}_w$, whose entries are i.i.d. zero-mean with variance $\frac{1}{n_T}$, is pre- and post-multiplied by the square root of deterministic matrices, $\mathbf{\Theta}_T$ and $\mathbf{\Theta}_R$, whose entries represent, respectively, the correlation between the transmit antennas and between the receive antennas:

$$\mathbf{H} = \mathbf{\Theta}_{\mathrm{R}}^{1/2} \mathbf{H}_w \mathbf{\Theta}_{\mathrm{T}}^{1/2} \tag{5.42}$$

Implied by this model is that the correlation between two transmit antennas is the same regardless of the receive antenna at which the observation is made and vice versa. The validity of this model has been confirmed by a number of experimental measurements conducted in various scenarios [48–54].

With full CSI at the transmitter, the asymptotic capacity is [55]

$$\mathcal{C}(\mathsf{SNR}) = \beta \int_0^\infty (\log(\mathsf{SNR}\,\nu\lambda))^+ dG(\lambda) \tag{5.43}$$

where ν satisfies

$$\int_0^\infty \left(\nu - \frac{1}{\mathsf{SNR}\,\lambda}\right)^+ dG(\lambda) = 1 \tag{5.44}$$

with $G(\cdot)$ the asymptotic spectrum of $\mathbf{H}^\dagger\mathbf{H}$ whose η transform can be derived using Theorem 4.7 and Lemma 4.2. Invoking Theorem 4.9, the capacity in (5.43) can be evaluated as follows.

Theorem 5.3 ([56]). *Let Λ_{R} and Λ_{T} be independent random variables whose distributions are the asymptotic spectra of the full-rank matrices $\mathbf{\Theta}_{\mathrm{R}}$ and $\mathbf{\Theta}_{\mathrm{T}}$ respectively. Further define*

$$\Lambda_1 = \begin{cases} \Lambda_{\mathrm{T}} & \beta < 1 \\ \Lambda_{\mathrm{R}} & \beta > 1 \end{cases} \qquad \Lambda_2 = \begin{cases} \Lambda_{\mathrm{R}} & \beta < 1 \\ \Lambda_{\mathrm{T}} & \beta > 1 \end{cases} \tag{5.45}$$

and let κ be the infimum (excluding any mass point at zero) of the support of the asymptotic spectrum of $\mathbf{H}^\dagger\mathbf{H}$. For

$$\mathsf{SNR} \geq \frac{1}{\kappa} - \delta\mathbb{E}\left[\frac{1}{\Lambda_1}\right] \tag{5.46}$$

with δ satisfying

$$\eta_{\Lambda_2}(\delta) = 1 - \min\left\{\beta, \frac{1}{\beta}\right\},$$

the asymptotic capacity of a channel with separable correlations and full CSI at the transmitter is

$$\mathcal{C}(\mathsf{SNR}) = \begin{cases} \beta\,\mathbb{E}\left[\log\dfrac{\Lambda_{\mathrm{T}}}{e\vartheta}\right] + \mathcal{V}_{\Lambda_{\mathrm{R}}}(\vartheta) + \beta\log\left(\mathsf{SNR} + \vartheta\mathbb{E}\left[\dfrac{1}{\Lambda_{\mathrm{T}}}\right]\right) & \beta < 1 \\ \mathbb{E}\left[\log\dfrac{\Lambda_{\mathrm{R}}}{\alpha e}\right] + \beta\,\mathcal{V}_{\Lambda_{\mathrm{T}}}(\alpha) + \log\left(\mathsf{SNR} + \alpha\mathbb{E}\left[\dfrac{1}{\Lambda_{\mathrm{R}}}\right]\right) & \beta > 1 \end{cases}$$

with α and ϑ the solutions to

$$\eta_{\Lambda_T}(\alpha) = 1 - \frac{1}{\beta} \qquad \eta_{\Lambda_R}(\vartheta) = 1 - \beta.$$

No asymptotic characterization of the capacity with full CSI at the transmitter is known for $\beta = 1$ and arbitrary SNR.

When the correlation is present only at either the transmit or receive ends of the link, the solutions in Theorem 5.3 sometimes become explicit:

Corollary 5.4. *With correlation at the end of the link with the fewest antennas, the capacity per antenna with full CSI at the transmitter converges to*

$$\mathcal{C} = \begin{cases} \beta \mathbb{E}\left[\log \frac{\Lambda_T}{e}\right] + \log \frac{1}{1-\beta} + \beta \log\left(\mathsf{SNR}\frac{1-\beta}{\beta} + \mathbb{E}\left[\frac{1}{\Lambda_T}\right]\right) & \begin{matrix}\beta < 1 \\ \Lambda_R = 1\end{matrix} \\ E\left[\log \frac{\Lambda_R}{e}\right] - \beta \log \frac{\beta - 1}{\beta} + \log\left(\mathsf{SNR}(\beta - 1) + \mathbb{E}\left[\frac{1}{\Lambda_R}\right]\right) & \begin{matrix}\beta > 1 \\ \Lambda_T = 1\,.\end{matrix} \end{cases}$$

Finally if all antennas are assumed uncorrelated — a single-user multi-antenna channel with no correlation (i.e $\mathbf{\Theta}_R = \mathbf{\Theta}_T = \mathbf{I}$) is commonly refereed to as canonical channel — the capacity per antenna with full CSI at the transmitter converges to:

Theorem 5.5 ([56]). *For*

$$\mathsf{SNR} \geq \frac{2\min\{1, \beta^{3/2}\}}{|1 - \sqrt{\beta}||1 - \beta|} \tag{5.47}$$

the capacity of the canonical channel with full CSI at the transmitter converges a.s. to

$$\mathcal{C}(\mathsf{SNR}) = \begin{cases} \beta \log\left(\frac{\mathsf{SNR}}{\beta} + \frac{1}{1-\beta}\right) + (1-\beta)\log\frac{1}{1-\beta} - \beta \log e & \beta < 1 \\ \log\left(\beta\,\mathsf{SNR} + \frac{\beta}{\beta - 1}\right) + (\beta - 1)\log \frac{\beta}{\beta - 1} - \log e & \beta > 1\,. \end{cases}$$

With statistical CSI at the transmitter, achieving capacity requires that the eigenvectors of the input covariance, $\mathbf{\Phi}$, coincide with those of $\mathbf{\Theta}_T$ [57, 58]. Consequently, denoting by $\mathbf{\Lambda}_T$ and $\mathbf{\Lambda}_R$ the diagonal eigenvalue matrices of $\mathbf{\Theta}_T$ and $\mathbf{\Theta}_R$, respectively, we have that

$$\mathcal{C}(\beta, \mathsf{SNR}) = \frac{1}{N} \log\det\left(\mathbf{I} + \mathsf{SNR}\,\mathbf{\Lambda}_R^{1/2}\mathbf{H}_w\mathbf{\Lambda}_T^{1/2}\mathbf{P}\mathbf{\Lambda}_T^{1/2}\mathbf{H}_w^\dagger\mathbf{\Lambda}_R^{1/2}\right)$$

where **P** is the capacity-achieving power allocation [44]. Applying Theorem 4.7, we obtain:

Theorem 5.6 ([21]). *The capacity of a Rayleigh-faded channel with separable transmit and receive correlation matrices* $\boldsymbol{\Theta}_{\mathrm{T}}$ *and* $\boldsymbol{\Theta}_{\mathrm{R}}$ *and statistical CSI at the transmitter converges to*

$$\mathcal{C}(\beta, \mathsf{SNR}) = \beta E\left[\log(1 + \mathsf{SNR}\, \Lambda \Gamma(\mathsf{SNR}))\right] + E\left[\log(1 + \mathsf{SNR}\, \Lambda_{\mathsf{R}} \Upsilon(\mathsf{SNR})\right] - \beta\, \mathsf{SNR}\, \Gamma(\mathsf{SNR}) \Upsilon(\mathsf{SNR}) \log e \tag{5.48}$$

where

$$\Gamma(\mathsf{SNR}) = \frac{1}{\beta} E\left[\frac{\Lambda_{\mathsf{R}}}{1 + \mathsf{SNR}\, \Lambda_{\mathsf{R}} \Upsilon(\mathsf{SNR})}\right] \tag{5.49}$$

$$\Upsilon(\mathsf{SNR}) = E\left[\frac{\Lambda}{1 + \mathsf{SNR}\, \Lambda \Gamma(\mathsf{SNR})}\right] \tag{5.50}$$

with expectation over Λ *and* Λ_{R} *whose distributions are given by the asymptotic empirical eigenvalue distributions of* $\boldsymbol{\Lambda}_{\mathrm{T}}\mathbf{P}$ *and* $\boldsymbol{\Theta}_{\mathrm{R}}$*, respectively.*

If the input is isotropic, the achievable mutual information is easily found from the foregoing result.

Corollary 5.7 ([59]). *Consider a channel defined as in Theorem 5.6 and an isotropic input. Expression (5.48) yields the mutual information with the distribution of* Λ *given by the asymptotic empirical eigenvalue distribution of* $\boldsymbol{\Theta}_{\mathrm{T}}$*.*

This corollary is illustrated in Fig. 2, which depicts the mutual information (bits/s/Hz) achieved by an isotropic input for a wide range of SNR. The channel is Rayleigh-faded with $n_{\mathrm{T}} = 4$ correlated antennas and $n_{\mathrm{R}} = 2$ uncorrelated antennas. The correlation between the ith and jth transmit antennas is

$$(\boldsymbol{\Theta}_{\mathrm{T}})_{i,j} = e^{-0.05 d^2 (i-j)^2} \tag{5.51}$$

which corresponds to a uniform linear array with antenna separation d (wavelengths) exposed to a broadside Gaussian azimuth angular spectrum with a 2° root-mean-square spread [60]. Such angular spread is typical of an elevated base station in rural or suburban areas. The solid lines depict the analytical solution obtained by applying Theorem 5.6 with $\mathbf{P} = \mathbf{I}$ and $\boldsymbol{\Theta}_{\mathrm{R}} = \mathbf{I}$ and with the expectations over Λ replaced with arithmetic averages over the eigenvalues of $\boldsymbol{\Theta}_{\mathrm{T}}$. The circles, in turn, show the result of Monte-Carlo simulations. Notice the excellent agreement even for such small numbers of antennas.

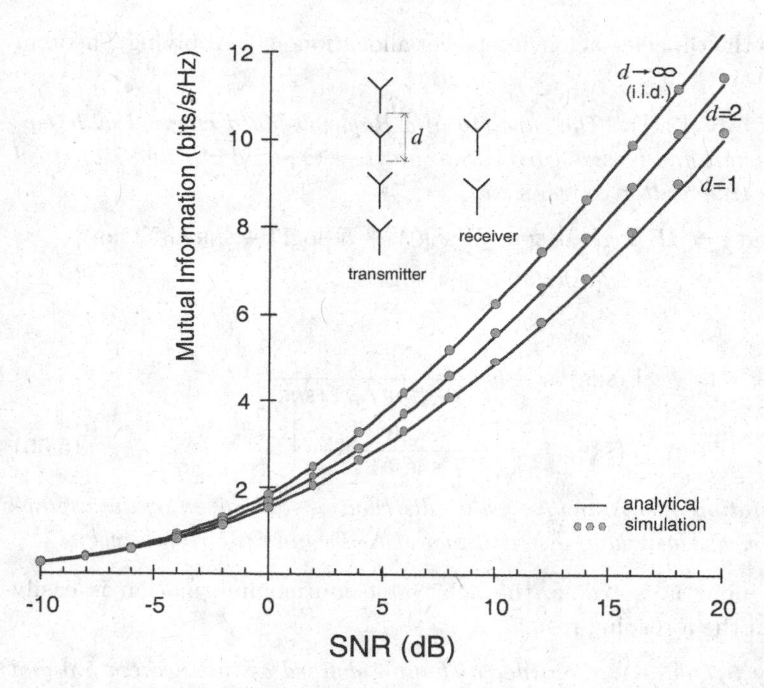

Fig. 2. Mutual information achieved by an isotropic input on a Rayleigh-faded channel with $n_{\mathrm{T}} = 4$ and $n_{\mathrm{R}} = 2$. The transmitter is a uniform linear array whose antenna correlation is given by (5.51) where d is the spacing (wavelengths) between adjacent antennas. The receive antennas are uncorrelated.

A Ricean term can be incorporated in the model (5.42) through an additional deterministic matrix $\mathbf{H}_0$ containing unit-magnitude entries [61–63]. With proper weighting of the random and deterministic matrices, the model particularizes to

$$\mathbf{y} = \left(\sqrt{\frac{1}{\mathsf{K}+1}} \boldsymbol{\Theta}_{\mathsf{R}}^{1/2} \mathbf{H}_w \boldsymbol{\Theta}_{\mathsf{T}}^{1/2} + \sqrt{\frac{\mathsf{K}}{\mathsf{K}+1}} \mathbf{H}_0 \right) \mathbf{x} + \mathbf{n} \tag{5.52}$$

with $\mathbf{H}_w$ an i.i.d. $\mathcal{N}(0,1)$ matrix and with the Ricean K-factor quantifying the ratio between the deterministic (unfaded) and the random (faded) energies [64].

If we assume that $\mathbf{H}_0$ has rank r where $r > 1$ but such that

$$\lim_{N\to\infty} \frac{r}{N} = 0 \tag{5.53}$$

then all foregoing results can be extended to the ricean channel by simply replacing Λ_R and Λ_T with independent random variables whose distributions are the asymptotic spectra of the full-rank matrices $\sqrt{\frac{1}{K+1}}\boldsymbol{\Theta}_R$ and $\sqrt{\frac{1}{K+1}}\boldsymbol{\Theta}_T$ respectively.

5.4. *Non-separable correlation model*

While the separable correlation model is relatively simple and analytically appealing, it also has clear limitations, particularly in terms of representing indoor propagation environments [65]. Also, it does not accommodate diversity mechanisms such as polarization[d] and radiation pattern diversity[e] that are becoming increasingly popular as they enable more compact arrays. The use of different polarizations and/or radiation patterns creates correlation structures that cannot be represented through the separable model.

A broader range of correlations can be encompassed, if we model the channel as

$$\mathbf{H} = \mathbf{U}_R \tilde{\mathbf{H}} \mathbf{U}_T^\dagger \tag{5.54}$$

where $\mathbf{U}_R$ and $\mathbf{U}_T$ are unitary while the entries of $\tilde{\mathbf{H}}$ are independent zero-mean Gaussian. This model is advocated and experimentally supported in [68] and its capacity is characterized asymptotically in [21]. For the more restrictive case where $\mathbf{U}_R$ and $\mathbf{U}_T$ are Fourier matrices, the model (5.54) was proposed earlier in [69].

Since the spectra of $\mathbf{H}$ and $\tilde{\mathbf{H}}$ coincide, every result derived for matrices with independent non-identically distributed entries (cf. Theorems 4.12–4.17) apply immediately to $\mathbf{H}$.

As it turns out, the asymptotic spectral efficiency of $\tilde{\mathbf{H}}$ is fully characterized by the variances of its entries, which we assemble in a matrix $\mathbf{G}$ such that $\mathsf{G}_{i,j} = n_T \mathbb{E}[|\mathsf{H}_{i,j}|^2]$ with

$$\sum_{ij} \mathsf{G}_{i,j} = n_T n_R \,. \tag{5.55}$$

---

[d] Polarization diversity: Antennas with orthogonal polarizations are used to ensure low levels of correlation with minimum or no antenna spacing [63, 66] and to make the communication link robust to polarization rotations in the channel [67].

[e] Pattern diversity: Antennas with different radiation patterns or with rotated versions of the same pattern are used to discriminate different multipath components and reduce correlation.

Invoking Definition 4.13, we introduce the *variance profile* of $\tilde{\mathbf{H}}$, which maps the entries of $\mathbf{G}$ onto a two-dimensional piece-wise constant function

$$\mathcal{G}^{(n_\mathrm{R})}(r,t) = \mathsf{G}_{i,j} \qquad \frac{i}{n_\mathrm{R}} \le r < \frac{i+1}{n_\mathrm{R}}, \quad \frac{j}{n_\mathrm{T}} \le t < \frac{j+1}{n_\mathrm{T}} \tag{5.56}$$

supported on $r, t \in [0,1]$. We can interpret r and t as normalized receive and transmit antenna indices. It is assumed that, as the number of antennas grows, $\mathcal{G}^{(n_\mathrm{R})}(r,t)$ converges uniformly to the *asymptotic variance profile*, $\mathcal{G}(r,t)$. The normalization condition in (5.55) implies that

$$\mathbb{E}[\mathcal{G}(\mathsf{R},\mathsf{T})] = 1 \tag{5.57}$$

with R and T independent random variables uniform on $[0,1]$.

With full CSI at the transmitter, the asymptotic capacity is given by (5.43) and (5.44) with $G(\cdot)$ representing the asymptotic spectrum of $\mathbf{H}^\dagger\mathbf{H}$. Using Theorems 4.17, an explicit expression for $\mathcal{C}(\mathsf{SNR})$ can be obtained for sufficiently high SNR.

With statistical CSI at the transmitter, the eigenvectors of the capacity-achieving input covariance coincide with the columns of $\mathbf{U}_\mathrm{T}$ in (5.54) [70, 71]. Consequently, the capacity is given by:

$$\mathcal{C}(\beta,\mathsf{SNR}) = \lim_{N\to\infty} \frac{1}{N} \log\det\left(\mathbf{I} + \tilde{\mathbf{H}}\mathbf{P}\tilde{\mathbf{H}}^\dagger\right). \tag{5.58}$$

Denote by $\mathcal{P}(t,\mathsf{SNR})$ the asymptotic power profile of the capacity achieving power allocation at each SNR, in order to characterize (5.58), we invoke Theorem 4.16 to obtain the following.

Theorem 5.8 ([21]). *Consider the channel* $\mathbf{H} = \mathbf{U}_\mathrm{R}\tilde{\mathbf{H}}\mathbf{U}_\mathrm{T}^\dagger$ *where* $\mathbf{U}_\mathrm{R}$ *and* $\mathbf{U}_\mathrm{T}$ *are unitary while the entries of* $\tilde{\mathbf{H}}$ *are zero-mean Gaussian and independent. Denote by* $\mathcal{G}(r,t)$ *the asymptotic variance profile of* $\tilde{\mathbf{H}}$. *With statistical CSI at the transmitter, the asymptotic capacity is*

$$\begin{aligned}\mathcal{C}(\beta,\mathsf{SNR}) = {} & \beta\,\mathbb{E}\left[\log(1+\mathsf{SNR}\,\mathbb{E}\left[\mathcal{G}(\mathsf{R},\mathsf{T})\mathcal{P}(\mathsf{T},\mathsf{SNR})\Gamma(\mathsf{R},\mathsf{SNR})\right|\mathsf{T}])\right] \\ & + \mathbb{E}\left[\log(1+\mathbb{E}[\mathcal{G}(\mathsf{R},\mathsf{T})\mathcal{P}(\mathsf{T},\mathsf{SNR})\Upsilon(\mathsf{T},\mathsf{SNR})|\mathsf{R}])\right] \\ & - \beta\,\mathbb{E}\left[\mathcal{G}(\mathsf{R},\mathsf{T})\mathcal{P}(\mathsf{T},\mathsf{SNR})\Gamma(\mathsf{R},\mathsf{SNR})\Upsilon(\mathsf{T},\mathsf{SNR})\right]\log e\end{aligned}$$

with expectation over the independent random variables R *and* T *uniform on* $[0,1]$ *and with*

$$\begin{aligned}\beta\,\Gamma(r,\mathsf{SNR}) &= \frac{1}{1+\mathbb{E}[\mathcal{G}(r,\mathsf{T})\mathcal{P}(\mathsf{T},\mathsf{SNR})\Upsilon(\mathsf{T},\mathsf{SNR})]} \\ \Upsilon(t,\mathsf{SNR}) &= \frac{\mathsf{SNR}}{1+\mathsf{SNR}\,\mathbb{E}\left[\mathcal{G}(\mathsf{R},t)\mathcal{P}(t,\mathsf{SNR})\Gamma(\mathsf{R},\mathsf{SNR})\right]}.\end{aligned}$$

If there is no correlation but antennas with different polarizations are used, the entries of $\mathbf{H}$ are no longer identically distributed because of the different power transfer between co-polarized and differently polarized antennas. In this case, we can model the channel matrix as

$$\mathbf{H} = \mathbf{A} \circ \mathbf{H}_w \tag{5.59}$$

where $\circ$ indicates Hadamard (element-wise) multiplication, $\mathbf{H}_w$ is composed of zero-mean i.i.d. Gaussian entries with variance $\frac{1}{n_T}$ and $\mathbf{A}$ is a deterministic matrix containing the square-root of the second-order moment of each entry of $\mathbf{H}$, which is given by the relative polarization of the corresponding antenna pair. If all antennas are co-polar, then every entry of $\mathbf{A}$ equals 1.

The asymptotic capacity with full CSI at the transmitter can be found, for sufficiently high SNR, by invoking Theorem 4.17.

Since the entries of $\mathbf{H}$ are independent, the input covariance that achieves capacity with statistical CSI is diagonal [70, 71]. The corresponding asymptotic capacity per antenna equals the one given in Theorem 5.8 with $\mathcal{G}(r,t)$ the asymptotic variance profile of $\mathbf{H}$. Furthermore, these solutions do not require that the entries of $\mathbf{H}$ be Gaussian but only that their variances be uniformly bounded.

A common structure for $\mathbf{A}$, arising when the transmit and receive arrays have an equal number of antennas on each polarization, is that of a doubly-regular form (cf. Definition 4.11). For such channels, the capacity-achieving input is not only diagonal but isotropic and, applying Theorem 4.12, the capacity admits an explicit form.

Theorem 5.9. *Consider a channel $\mathbf{H} = \mathbf{A} \circ \mathbf{H}_w$ where the entries of $\mathbf{A}$ are deterministic and nonnegative while those of $\mathbf{H}_w$ are zero-mean and independent, with variance $\frac{1}{n_T}$ but not necessarily identically distributed. If $\mathbf{A}$ is doubly-regular (cf. Definition 4.11), the asymptotic capacity per antenna, with full CSI or with statistical CSI at the transmitter, coincides with that of the canonical channel, given respectively in Theorem 5.5 and in Eq. (4.10) with in the latter $\gamma = \frac{\mathsf{SNR}}{\beta}$.*

A very practical example of the applicability of the above result is given by the following wireless channel.

Example 5.10. Consider the wireless channel as in Fig. 3 where each transmitter and receiver have antennas split between two orthogonal polarizations. Denoting by σ the gain between copolar antennas different from

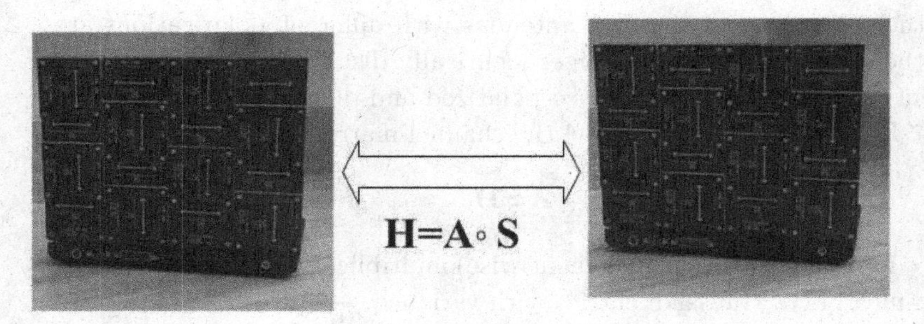

Fig. 3. Laptop computers equipped with a 16-antenna planar array. Two orthogonal polarizations used.

gain between crosspolar antennas, χ, we can model the channel matrix as in (5.59), where $\mathbf{P} = \mathbf{A} \circ \mathbf{A}$ equals:

$$\mathbf{P} = \begin{bmatrix} \sigma & \chi & \sigma & \chi & \cdots \\ \chi & \sigma & \chi & \sigma & \cdots \\ \sigma & \chi & \sigma & \chi & \cdots \\ \vdots & \vdots & \vdots & \vdots & \ddots \end{bmatrix} \tag{5.60}$$

which is asymptotically mean doubly regular.

Again the zero-mean multi-antenna channel model analyzed thus far can be made Ricean by incorporating an additional deterministic component $\bar{\mathbf{H}}$ [61–63] which leads to the following general model

$$\mathbf{y} = \left(\sqrt{\frac{1}{K+1}} \mathbf{H} + \sqrt{\frac{K}{K+1}} \bar{\mathbf{H}} \right) \mathbf{x} + \mathbf{n} \tag{5.61}$$

with the scalar Ricean factor K quantifying the ratio between the Frobenius norm of the deterministic (unfaded) component and the expected Frobenius norm of the random (faded) component. Considered individually, each (i, j)th channel entry has a Ricean factor given by

$$K \frac{|\bar{\mathsf{H}}_{i,j}|^2}{\mathbb{E}[|\mathsf{H}_{i,j}|^2]}.$$

Using Lemma 2.6 in [23] the next result follows straightforwardly.

Theorem 5.11. *Consider a channel with a Ricean term whose rank is finite. The asymptotic capacity per antenna,* $\mathcal{C}^{\text{rice}}(\beta, \text{SNR})$*, equals the corresponding asymptotic capacity per antenna in the absence of the Ricean* *component, $\mathcal{C}(\beta, \mathsf{SNR})$, with a simple SNR penalty:*

$$\mathcal{C}^{\text{rice}}(\beta, \mathsf{SNR}) = \mathcal{C}\left(\beta, \frac{\mathsf{SNR}}{K+1}\right) \tag{5.62}$$

Note that, while the value of the capacity depends on the degree of CSI available at the transmitter, (5.62) holds regardless.

5.5. *Non-ergodic channels*

The results on large random matrices surveyed in Section 4 show that the mutual information per receive antenna converges a.s. to its expectation as the number of antennas goes to infinity (with a given ratio of transmit to receive antennas). Thus, as the number of antennas grows, a self-averaging mechanism hardens the mutual information to its expected value. However, the non-normalized mutual information still suffers random fluctuations that, although small with respect to the mean, are of vital interest in the study of the outage capacity.

An interesting property of the distribution of the non-normalized mutual information in (3.18) is the fact that, for many of the multi-antenna channels of interest, it can be approximated as Gaussian as the number of antennas grows. A number of authors have explored this property. Arguments supporting the normality of the c.d.f (cumulative distribution function) of the mutual information for large numbers of antennas were given in [29, 72–74].[f] Ref. [72] used the replica method from statistical physics (which has yet to find a rigorous justification), [73] showed the asymptotic normality only in the asymptotic regimes of low and high signal-to-noise ratios, while in [74], the normality of the outage capacity is proved for the canonical channel using [17]. Theorem 4.19 proves the asymptotic normality of the unnormalized mutual information for arbitrary signal-to-noise ratios and fading distributions, allowing for correlation between the antennas at either transmitter or receiver. Theorem 4.20 — a proof of such theorem can be found in [29] — provides succinct expressions for the asymptotic mean and variance of the mutual information in terms of the η and Shannon transforms of the correlation matrix. Using Theorem 4.20 we can get an extremely accurate approximation of the cumulative distribution of (3.18) with an arbitrary number of transmit and receive antennas. More specifically we have that the cumulative distribution of the unnormalized mutual

[f] For additional references, see [3].

information of a MIMO channel with correlation at the transmitter for arbitrary signal-to-noise ratios and fading distributions, is well approximated by a Gaussian distribution with mean, μ and varaince σ^2 given by

$$\mu = \frac{1}{N}\sum_{j=1}^{n_{\mathrm{T}}} \log\left(1 + \mathsf{SNR}\lambda_j(\mathbf{T})\,\eta\right) - \log\eta + (\eta - 1)\log e \tag{5.63}$$

$$\sigma^2 = -\log\left(1 - \beta\frac{1}{n_{\mathrm{T}}}\sum_{j=1}^{n_{\mathrm{T}}}\left[\left(\frac{\lambda_j(\mathbf{T})\,\mathsf{SNR}\,\eta}{1+\lambda_j(\mathbf{T})\,\mathsf{SNR}\,\eta}\right)^2\right]\right). \tag{5.64}$$

In order to illustrate the power of this result with some examples, we will consider correlated MIMO channels with a transmit correlation matrix $\mathbf{\Theta}_{\mathrm{T}}$ such that

$$(\mathbf{\Theta}_{\mathrm{T}})_{i,j} = e^{-0.8(i-j)^2} \tag{5.65}$$

which is a typical structure of an elevated base station in suburbia. The receive antennas are uncorrelated. For the examples, we will compare the cumulative distributions of the unnormalized mutual information of such channel with a Gaussian distribution whose mean and variance are given in (5.63) and (5.64). Figures 4 and 5 compare the 10% point in the cumulative

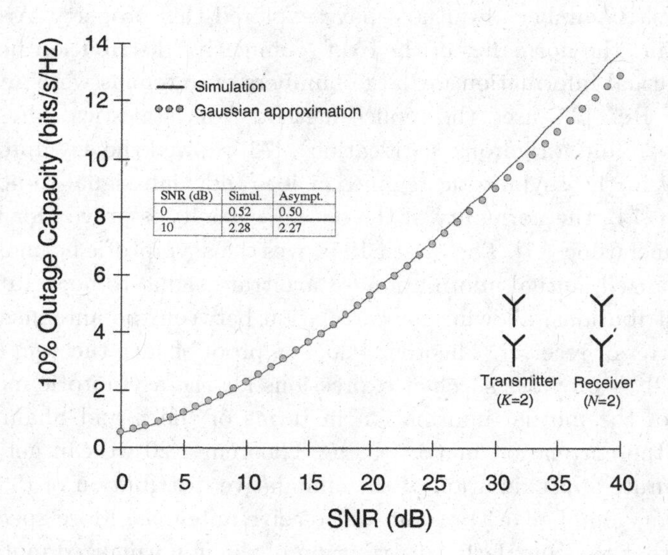

SNR (dB)	Simul.	Asympt.
0	0.52	0.50
10	2.28	2.27

Fig. 4. 10%-outage capacity for a Rayleigh-faded channel with $n_{\mathrm{T}} = n_{\mathrm{R}} = 2$. The transmit antennas are correlated as per (5.65) while the receive antennas are uncorrelated. Solid line indicates the corresponding limiting Gaussian distribution.

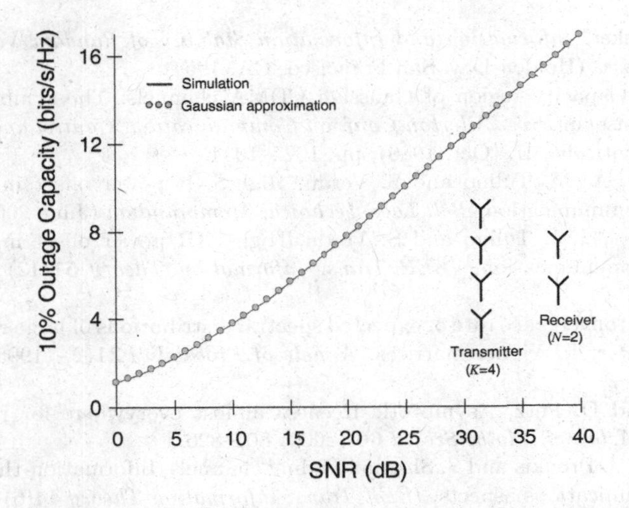

Fig. 5. 10%-outage capacity for a Rayleigh-faded channel with $n_{\mathrm{T}} = 4$ and $n_{\mathrm{R}} = 2$. The transmit antennas are correlated as per (5.65) while the receive antennas are uncorrelated. Solid line indicates the corresponding limiting Gaussian distribution.

distribution of the mutual information for SNR between 0 and 40 dB for $n_{\mathrm{R}} = 2$ and different number of transmit antennas. The solid line indicates the simulation while the circles indicate the Gaussian distribution. Notice the remarkable agreement despite having such a small number of antennas.

For channels with both transmit and receive correlation, the characteristic function found through the replica method yields to the expression of $\mathbb{E}[\Delta^2]$ given in [72].

References

1. S. Verdú, Random matrices in wireless communication, proposal to the National Science Foundation (Feb. 1999).
2. S. Verdú, Large random matrices and wireless communications, *2002 MSRI Information Theory Workshop* (Feb 25–Mar 1, 2002).
3. A. M. Tulino and S. Verdú, *Random Matrix Theory and Wireless Communications*, Foundations and Trends in Communications and Information Theory, Volume 1, Issue 1 (Now Publishers Inc., 2004).
4. S. Verdú, *Multiuser Detection* (Cambridge University Press, Cambridge, UK, 1998).
5. S. Verdú and S. Shamai, Spectral efficiency of CDMA with random spreading, *IEEE Trans. Information Theory* **45**(2) (1999) 622–640.
6. D. Tse and S. Hanly, Linear multiuser receivers: Effective interference, effective bandwidth and user capacity, *IEEE Trans. Information Theory* **45**(2) (1999) 641–657.

7. M. S. Pinsker, *Information and Information Stability of Random Variables and Processes* (Holden-Day, San Francisco, CA, 1964).
8. S. Verdú, Capacity region of Gaussian CDMA channels: The symbol synchronous case, in *Proc. Allerton Conf. on Communication, Control and Computing*, Monticello, IL (Oct. 1986), pp. 1025–1034.
9. A. Lozano, A. M. Tulino and S. Verdú, High-SNR power offset in multiantenna communication, *Bell Labs Technical Memorandum* (June 2004).
10. A. Lozano, A. M. Tulino and S. Verdú, High-SNR power offset in multiantenna communication, *IEEE Trans. Information Theory* **51**(12) (2005) 4134–4151.
11. Z. D. Bai, Convergence rate of expected spectral distributions of large random matrices. Part I: Wigner matrices, *Annals of Probability* **21**(2) (1993) 625–648.
12. F. Hiai and D. Petz, Asymptotic freeness almost everywhere for random matrices, *Acta Sci. Math. Szeged* **66** (2000) 801–826.
13. E. Biglieri, J. Proakis and S. Shamai, Fading channels: Information-theoretic and communications aspects, *IEEE Trans. Information Theory* **44**(6) (1998) 2619–2692.
14. I. Csiszár and J. Körner, *Information Theory: Coding Theorems for Discrete Memoryless Systems* (Academic, New York, 1981).
15. G. J. Foschini, Layered space-time architecture for wireless communication in a fading environment when using multi-element antennas, *Bell Labs Technical Journal* **1** (1996) 41–59.
16. E. Telatar, Capacity of multi-antenna Gaussian channels, *Euro. Trans. Telecommunications* **10**(6) (1999) 585–595.
17. Z. D. Bai and J. W. Silverstein, CLT of linear spectral statistics of large dimensional sample covariance matrices, *Annals of Probability* **32**(1A) (2004) 553–605.
18. T. J. Stieltjes, Recherches sur les fractions continues, *Annales de la Faculte des Sciences de Toulouse* **8**(9) (1894) (1895) no. A (J), pp. 1–47 (1–122).
19. V. A. Marčenko and L. A. Pastur, Distributions of eigenvalues for some sets of random matrices, *Math. USSR-Sbornik* **1** (1967) 457–483.
20. J. W. Silverstein and Z. D. Bai, On the empirical distribution of eigenvalues of a class of large dimensional random matrices, *J. Multivariate Analysis* **54** (1995) 175–192.
21. A. M. Tulino, A. Lozano and S. Verdú, Impact of correlation on the capacity of multi-antenna channels, *IEEE Trans. Information Theory* **51**(7) (2005) 2491–2509.
22. A. Lozano, A. M. Tulino and S. Verdú, High-SNR power offset in multiantenna communication, in *Proc. IEEE Int. Symp. on Information Theory (ISIT'04)*, Chicago, IL (June 2004).
23. Z. D. Bai, Methodologies in spectral analysis of large dimensional random matrices, *Statistica Sinica* **9**(3) (1999) 611–661.
24. V. L. Girko, *Theory of Random Determinants* (Kluwer Academic Publishers, Dordrecht, 1990).

25. A. Guionnet and O. Zeitouni, Concentration of the spectral measure for large matrices, *Electronic Communications in Probability* **5** (2000) 119–136.
26. D. Shlyankhtenko, Random Gaussian band matrices and freeness with amalgamation, *Int. Math. Res. Note* **20** (1996) 1013–1025.
27. L. Li, A. M. Tulino and S. Verdú, Spectral efficiency of multicarrier CDMA, *IEEE Trans. Information Theory* **51**(2) (2005) 479–505.
28. Z. D. Bai and J. W. Silverstein, Exact separation of eigenvalues of large dimensional sample covariance matrices, *Annals of Probability* **27**(3) (1999) 1536–1555.
29. A. M. Tulino and S. Verdú, Asymptotic outage capacity of multiantenna channels, in *Proc. IEEE Int. Conf. Acoustics, Speech and Signal Processing (ICASSP'05)*, Philadelphia, PA, USA (Mar. 2005).
30. S. Shamai and S. Verdú, The effect of frequency-flat fading on the spectral efficiency of CDMA, *IEEE Trans. Information Theory* **47**(4) (2001) 1302–1327.
31. J. M. Chaufray, W. Hachem and P. Loubaton, Asymptotic analysis of optimum and sub-optimum CDMA MMSE receivers, *Proc. IEEE Int. Symp. on Information Theory (ISIT'02)* (July 2002), p. 189.
32. L. Li, A. M. Tulino and S. Verdú, Design of reduced-rank MMSE multiuser detectors using random matrix methods, *IEEE Trans. Information Theory* **50**(6) (2004).
33. M. Debbah, W. Hachem, P. Loubaton and M. de Courville, MMSE analysis of certain large isometric random precoded systems, *IEEE Trans. Information Theory* **49**(5) (2003) 1293–1311.
34. R. Horn and C. Johnson, *Matrix Analysis* (Cambridge University Press, 1985).
35. G. Foschini and M. Gans, On limits of wireless communications in fading environment when using multiple antennas, *Wireless Personal Communications* **6**(6) (1998) 315–335.
36. S. N. Diggavi, N. Al-Dhahir, A. Stamoulis and A. R. Calderbank, Great expectations: The value of spatial diversity in wireless networks, *Proc. IEEE* **92**(2) (2004) 219–270.
37. A. Goldsmith, S. A. Jafar, N. Jindal and S. Vishwanath, Capacity limits of MIMO channels, *IEEE J. Selected Areas in Communications* **21**(5) (2003) 684–702.
38. D. Gesbert, M. Shafi, D. Shiu, P. J. Smith and A. Naguib, From theory to practice: An overview of MIMO space–time coded wireless systems, *J. Selected Areas in Communications* **21**(3) (2003) 281–302.
39. E. Biglieri and G. Taricco, Large-system analyses of multiple-antenna system capacities, *Journal of Communications and Networks* **5**(2) (2003) 57–64.
40. E. Biglieri and G. Taricco, Transmission and reception with multiple antennas: Theoretical foundations, submitted to *Foundations and Trends in Communications and Information Theory* (2004).
41. B. S. Tsybakov, The capacity of a memoryless Gaussian vector channel, *Problems of Information Transmission* **1** (1965) 18–29.

42. T. M. Cover and J. A. Thomas, *Elements of Information Theory* (John Wiley and Sons, Inc., 1991).
43. G. Raleigh and J. M. Cioffi, Spatio-temporal coding for wireless communications, *IEEE Trans. on Communications* **46**(3) (1998) 357–366.
44. A. M. Tulino, A. Lozano and S. Verdú, Power allocation in multi-antenna communication with statistical channel information at the transmitter, in *Proc. IEEE Int. Conf. on Personal, Indoor and Mobile Radio Communications (PIMRC'04)*, Barcelona, Catalonia, Spain (Sep. 2004).
45. D.-S. Shiu, G. J. Foschini, M. J. Gans and J. M. Kahn, Fading correlation and its effects on the capacity of multi-element antenna systems, *IEEE Trans. on Communications* **48**(3) (2000) 502–511.
46. D. Chizhik, F. R. Farrokhi, J. Ling and A. Lozano, Effect of antenna separation on the capacity of BLAST in correlated channels, *IEEE Communications Letters* **4**(11) (2000) 337–339.
47. K. I. Pedersen, J. B. Andersen, J. P. Kermoal and P. E. Mogensen, A stochastic multiple-input multiple-output radio channel model for evaluations of space-time coding algorithms, in *Proc. IEEE Vehicular Technology Conf. (VTC'2000 Fall)* (Sep. 2000), pp. 893–897.
48. C. C. Martin, J. H. Winters and N. R. Sollenberger, Multiple-input multiple-output (MIMO) radio channel measurements, in *Proc. IEEE Vehic. Techn. Conf. (VTC'2000)*, Boston, MA, USA (Sep. 2000).
49. H. Xu, M. J. Gans, N. Amitay and R. A. Valenzuela, Experimental verification of MTMR system capacity in a controlled propagation environment, *Electronics Letters* **37** (2001).
50. J. P. Kermoal, L. Schumacher, P. E. Mogensen and K. I. Pedersen, Experimental investigation of correlation properties of MIMO radio channels for indoor picocell scenarios, in *Proc. IEEE Vehic. Tech. Conf. (VTC'2000 Fall)* (2000).
51. D. Chizhik, G. J. Foschini, M. J. Gans and R. A. Valenzuela, Propagation and capacities of multi-element transmit and receive antennas, in *Proc. 2001 IEEE AP-S Int. Symp. and USNC/URSI National Radio Science Meeting*, Boston, MA (July 2001).
52. J. Ling, D. Chizhik, P. W. Wolniansky, R. A. Valenzuela, N. Costa and K. Huber, Multiple transmit multiple receive (MTMR) capacity survey in Manhattan, *IEE Electronics Letters* **37**(16) (2001) 1041–1042.
53. D. Chizhik, J. Ling, P. Wolniansky, R. A. Valenzuela, N. Costa and K. Huber, Multiple-input multiple-output measurements and modelling in Manhattan, *IEEE J. Selected Areas in Communications* **21**(3) (2003) 321–331.
54. V. Erceg, P. Soma, D. S. Baum and A. Paulraj, Capacity obtained from multiple-input multiple-output channel measurements in fixed wireless environments at 2.5 GHz, *Int. Conf. on Commun. (ICC'02)*, New York City, NY (Apr. 2002).
55. C. Chuah, D. Tse, J. Kahn and R. Valenzuela, Capacity scaling in dual-antenna-array wireless systems, *IEEE Trans. Information Theory* **48**(3) (2002) 637–650.

56. A. M. Tulino, A. Lozano and S. Verdú, MIMO capacity with channel state information at the transmitter, in *Proc. IEEE Int. Symp. on Spread Spectrum Tech. and Applications (ISSSTA'04)* (Aug. 2004).
57. E. Visotsky and U. Madhow, Space-time transmit precoding with imperfect feedback, *IEEE Trans. Information Theory* **47** (2001) 2632–2639.
58. S. A. Jafar, S. Vishwanath and A. J. Goldsmith, Channel capacity and beamforming for multiple transmit and receive antennas with covariance feedback, in *Proc. IEEE Int. Conf. on Communications (ICC'01)*, Vol. 7 (2001), pp. 2266–2270.
59. A. M. Tulino, S. Verdú and A. Lozano, Capacity of antenna arrays with space, polarization and pattern diversity, in *Proc. 2003 IEEE Information Theory Workshop (ITW'03)* (Apr. 2003), pp. 324–327.
60. T.-S. Chu and L. J. Greenstein, A semiempirical representation of antenna diversity gain at cellular and PCS base stations, *IEEE Trans. on Communications* **45**(6) (1997) 644–656.
61. P. Driessen and G. J. Foschini, On the capacity formula for multiple-input multiple-output channels: A geometric interpretation, *IEEE Trans. on Communications* **47**(2) (1999) 173–176.
62. F. R. Farrokhi, G. J. Foschini, A. Lozano and R. A. Valenzuela, Link-optimal space-time processing with multiple transmit and receive antennas, *IEEE Communications Letters* **5**(3) (2001) 85–87.
63. P. Soma, D. S. Baum, V. Erceg, R. Krishnamoorthy and A. Paulraj, Analysis and modelling of multiple-input multiple-output (MIMO) radio channel based on outdoor measurements conducted at 2.5 GHz for fixed BWA applications, in *Proc. IEEE Int. Conf. on Communications (ICC'02)*, New York City, NY (28 Apr.–2 May 2002), pp. 272–276.
64. S. Rice, Mathematical analysis of random noise, *Bell System Technical Journal* **23** (1944) 282–332.
65. H. Ozcelik, M. Herdin, W. Weichselberger, G. Wallace and E. Bonek, Deficiencies of the Kronecker MIMO channel model, *IEE Electronic Letters* **39** (2003) 209–210.
66. W. C. Y. Lee and Y. S. Yeh, Polarization diversity system for mobile radio, *IEEE Trans. on Communications* **20**(5) (1972) 912–923.
67. S. A. Bergmann and H. W. Arnold, Polarization diversity in portable communications environment, *IEE Electronic Letters* **22**(11) (1986) 609–610.
68. W. Weichselberger, M. Herdin, H. Ozcelik and E. Bonek, Stochastic MIMO channel model with joint correlation of both link ends, *IEEE Trans. Wireless Communications* **5**(1) (2006) 90–100.
69. A. Sayeed, Deconstructing multi-antenna channels, *IEEE Trans. Signal Processing* **50**(10) (2002) 2563–2579.
70. A. M. Tulino, A. Lozano and S. Verdú, Capacity-achieving input covariance for single-user multi-antenna channels, *Bell Labs Tech. Memorandum* ITD-04-45193Y, also in *IEEE Trans. Wireless Communications* **5**(3) (2006) 662–671.

71. V. V. Veeravalli, Y. Liang and A. Sayeed, Correlated MIMO Rayleigh fading channels: Capacity, optimal signalling and asymptotics, *IEEE Trans. Information Theory* **51**(6) (2005) 2058–2072.
72. A. L. Moustakas, S. H. Simon and A. M. Sengupta, MIMO capacity through correlated channels in the presence of correlated interferers and noise: A (not so) large N analysis, *IEEE Trans. Information Theory* **49**(10) (2003) 2545–2561.
73. B. M. Hochwald, T. L. Marzetta and V. Tarokh, Multi-antenna channel hardening and its implications for rate feedback and scheduling, *IEEE Trans. Information Theory* **50**(9) (2004) 1893–1909.
74. M. Kamath, B. Hughes and Y. Xinying, Gaussian approximations for the capacity of MIMO Rayleigh fading channels, in *Asilomar Conf. on Signals, Systems and Computers* (Nov. 2002).

THE REPLICA METHOD IN MULTIUSER COMMUNICATIONS

Ralf R. Müller
Department of Electronics and Telecommunications
Norwegian University of Science and Technology
7491 Trondheim, Norway
E-mail: ralf@iet.ntnu.no

This review paper gives a tutorial overview of the usage of the replica method in multiuser communications. It introduces the self averaging principle, the free energy and other physical quantities and gives them a meaning in the context of multiuser communications. The technical issues of the replica methods are explained to a non-physics audience. An isomorphism between receiver metrics and the fundamental laws of physics is drawn. The overview is explained at the example of detection of code-division multiple-access with random signature sequences.

1. Introduction

Multiuser communication systems which are driven by Gaussian distributed signals can be fully characterized by the distribution of the singular values of the channel matrix in the large user limit. In digital communications, however, transmitted signals are chosen from finite, often binary, sets. In those cases, knowledge of the asymptotic spectrum of large random matrices is, in general, not sufficient to get valuable insight into the behavior of characteristic performance measures such as bit error probabilities and supported data rate. We will see that the quantized nature of signals gives rise to the totally unexpected occurrence of phase transitions in multiuser communications which can, by no means, be inferred from the asymptotic convergence of eigenvalue spectra of large random matrices.

In order to analyze and design large dimensional communication systems which cannot be described by eigenvalues and eigenvectors alone, but depend on statistics of the transmitted signal, e.g. minimum distances between signal points, a more powerful machinery than random matrix and free probability theory is needed. Such a machinery was developed in statistical physics for the analysis of some particular magnetic materials called spin glasses and is known as the *replica method* [1]. Additionally, the replica method is well-tailored to cope with receivers whose knowledge about channel and/or input statistics is impaired.

The replica method is able to reproduce many of the results which were found by means of random matrix and free probability theory, but the calculations based on the replica method are often much more involved. Moreover, it is still lacking mathematical rigor in certain respects. However, due to its success in explaining physical phenomena and its consistency with engineering results from random matrix and free probability theory, we can trust that its predictions in other engineering applications are correct. Nevertheless, we should always exercise particular care when interpreting new results based on the replica method. Establishing a rigorous mathematical basis for the replica method is a topic of current research in mathematics and theoretical physics.

2. Self Average

While random matrix theory and recently also free probability theory [2, 3] prove the (almost sure) convergence of some random variables to deterministic values in the large matrix limit, statistical physics does not always do so. It is considered a fundamental principle of statistical physics that there are microscopic and macroscopic variables. Microscopic variables are physical properties of microscopically small particles, e.g. the speed of a gas molecule or the spin of an electron. Macroscopic variables are physical properties of compound objects that contain many microscopic particles, e.g. the temperature or pressure of a gas, the radiation of a hot object, or the magnetic field of a piece of ferromagnetic material. From a physics point of view, it is clear which variables are macroscopic and which ones are microscopic. An explicit proof that a particular variable is *self-averaging*, i.e. it converges to a deterministic value in the large system limit, is a nice result, if it is found, but it is not particularly important to the physics community. When applying the replica method, systems are often only assumed to be self-averaging. The replica method itself must be seen as a tool to enable the calculation of macroscopic properties by averaging over the microscopic properties.

3. Free Energy

The second law of thermodynamics demands the entropy of any physical system with conserved energy to converge to its maximum as time evolves. If the system is described by a density $\mathrm{p}_X(x)$ of states $X \in \mathbb{R}$, this means that in the thermodynamic equilibrium the (differential) entropy

$$\mathrm{H}(X) = -\int \log \mathrm{p}_X(x)\, \mathrm{dP}_X(x) \tag{3.1}$$

is maximized while keeping the energy

$$\mathrm{E}(X) = \int ||x||\, \mathrm{dP}_X(x) \tag{3.2}$$

constant. Hereby, the energy function $||x||$ can be any measure which is uniformly bounded from below.

The density at thermodynamic equilibrium is easily shown by the method of Lagrange multipliers to be

$$\mathrm{p}_X(x) = \frac{\mathrm{e}^{-\frac{1}{T}||x||}}{\displaystyle\int_{-\infty}^{+\infty} \mathrm{e}^{-\frac{1}{T}||x||}\, \mathrm{d}x} \tag{3.3}$$

and called the Boltzmann distribution. The parameter T is called the temperature of the system and determined by (3.2). For a Euclidean energy measure, the Boltzmann distribution takes on the form of a Gaussian distribution which is well known in information theory to maximize entropy for given average signal power.

A helpful quantity in statistical mechanics is the (normalized) *free energy*[a] defined as

$$\mathrm{F}(X) \triangleq \mathrm{E}(X) - T\mathrm{H}(X) \tag{3.4}$$

$$= -T \log\left(\int_{-\infty}^{+\infty} \mathrm{e}^{-\frac{1}{T}||x||}\, \mathrm{d}x\right). \tag{3.5}$$

In the thermodynamic equilibrium, the entropy is maximized and the free energy is minimized since the energy is constant. The free energy normalized to the dimension of the system is a self averaging quantity.

As we will see in the next section, the task of receivers in digital communications is to minimize an energy function for a given received signal. In the terminology of statistical physics, they minimize the energy for a

[a]The *free* energy is not related to *free*ness in free probability theory.

given entropy. Formulating this optimization problem in Lagrangian form, we find that the free energy (3.4) is the object function to be minimized and the temperature of the system is a Lagrange multiplier. We conclude that, also from an engineering point of view, the free energy is a natural quantity to look at.

4. The Meaning of the Energy Function

In statistical physics, the free energy characterizes the energy of a system at given entropy via the introduction of a Lagrange multiplier which is called temperature (3.4). This establishes the usefulness of the free energy for information theoretic tasks like calculations of channel capacities. Moreover, the free energy is a tool to analyze various types of multiuser detectors. In fact, the free energy is such a powerful concept that it needs not any coding to be involved in the communication system to yield striking results. The only condition, it requires to be fulfilled, is the existence of macroscopic variables, microscopic random variables and the existence of an energy function. For communication systems, this requires, in practice, nothing more than their size growing above all bounds.

In physics, the energy function is determined by the fundamental forces of physics. It can represent kinetic energy, energy contained in electric, magnetic or nuclear fields. The broad applicability of the statistical mechanics approach to communication systems stems form the validity of (3.4) for any definition of the energy function. The energy function can be interpreted as the metric of a detector. Thus, any detector parameterized by a certain metric can be analyzed with the tools of statistical mechanics in the large system limit. There is no need that the performance measures of the detectors depend only on the eigenvalues of the channel matrix in the large system limits. However, there is a practical limit to the applicability of the statistical mechanics framework to the analysis of large communication systems: The analytical calculations required to solve the equations arising from (3.4) are not always feasible. The replica method was introduced to circumvent such difficulties. Some cases, however, have remained intractable until present time.

Consider a communication channel uniquely characterized by a conditional probability density $\mathrm{p}_{Y|X}(y,x)$ and a source uniquely characterized by a prior density $\mathrm{p}_X(x)$. Consider a detector for the output of this channel characterized by an assumed channel transition probability $\check{\mathrm{p}}_{Y|X}(y,x)$ and an assumed prior distribution $\check{\mathrm{p}}_X(x)$. Let the detector minimize some kind of cost function, e.g. bit error probability, subject to its hypotheses on the channel transition probability $\breve{p}_{Y|X}(y,x)$ and the prior distribution $\breve{p}_X(x)$. If the assumed distributions equal the true distributions, the detector is optimum with respect to its cost function. If the assumed distributions differ from the true ones, the detector is mismatched in some sense. The mismatch can arise from insufficient knowledge at the detector due to channel fluctuations or due to detector complexity. If the optimum detector requires an exhaustive search to solve an np-complete optimization, approximations to the true prior distribution often lead to suboptimal detectors with reduced complexity. Many popular detectors can be described within this framework.

The minimization of a cost function subject to some hypothesis on the channel transition probability and some hypothesis on the prior distribution defines a metric which is to be optimized. This metric corresponds to the energy function in thermodynamics and determines the distribution of the microscopic variables in the thermodynamic equilibrium. In analogy to (3.3), we find

$$\breve{p}_{X|Y}(x,y) = \frac{\mathrm{e}^{-\frac{1}{T}||x||}}{\displaystyle\int_{-\infty}^{+\infty} \mathrm{e}^{-\frac{1}{T}||x||}\,\mathrm{d}x} \tag{4.1}$$

where the dependency on y and the assumed prior distribution is implicit via the definition of the energy function $||\cdot||$. The energy function reflects the properties of the detector. Using Bayes' law, the appropriate energy function corresponding to particular hypotheses on the channel transition function and the prior distribution can be calculated via (4.1). While the energy function in statistical physics is uniquely defined by the fundamental forces of physics, the energy function in digital communications characterizes the algorithm run in the detector. Thus, every different algorithm potentially run in a detector uniquely defines the statistical physics of a corresponding imaginary toy universe where the natural forces of physics have been replaced by some imaginary alternatives characterizing a particular detection algorithm.

In order to study macroscopic properties of the system, we must calculate the free energy of the system. For that purpose, we make use of the self-averaging property of the thermodynamic equilibrium and (3.5):

$$\mathrm{F}(X) = \mathop{\mathrm{E}}_{Y} \mathrm{F}(X|Y) \tag{4.2}$$

$$= -T \int \log\left(\int \mathrm{e}^{-\frac{1}{T}||x||}\,\mathrm{d}x\right) \mathrm{dP}_Y(y)\,. \tag{4.3}$$

Note that, inside the logarithm, expectations are taken with respect to the assumed distribution via the definition of the energy function, while, outside the logarithm, expectations are taken with respect to the true distribution.

In the case of matched detection, i.e. the assumed distributions equal the true distributions, the argument of the logarithm in (4.3) becomes $\mathrm{p}_Y(y)$ up to a normalizing factor. Thus, the free energy becomes the (differential) entropy of Y up to a scaling factor and an additive constant.

Statistical mechanics provides an excellent framework to study not only matched, but also mismatched detection. The analysis of mismatched detection in large communication systems which is purely based on asymptotic properties of large random matrices and does not exploit the tools provided by statistical mechanics has been very limited so far. One exception is the asymptotic SINR of linear MMSE multiuser detectors with erroneous assumptions on the powers of interfering users in [4].

5. Replica Continuity

The explicit evaluation of the free energy turns out to be very complicated in many cases of interest. One major obstacle is the occurrence of the expectation of the logarithm of some function $f(\cdot)$ of a random variable Y

$$\mathop{\mathrm{E}}_{Y} \log f(Y) \,. \tag{5.1}$$

In order to circumvent this expectation which also appears frequently in information theory, the following identities are helpful

$$\log Y = \lim_{n\to 0} \frac{Y^n - 1}{n} \tag{5.2}$$

$$= \lim_{n\to 0} \frac{\partial}{\partial n} Y^n. \tag{5.3}$$

Under the assumption that limit and expectation can be interchanged, this gives

$$\mathop{\mathrm{E}}_{Y} \log f(Y) = \lim_{n\to 0} \frac{\partial}{\partial n} \mathop{\mathrm{E}}_{Y} [f(Y)]^n \tag{5.4}$$

$$= \lim_{n\to 0} \frac{\partial}{\partial n} \log \mathop{\mathrm{E}}_{Y} [f(Y)]^n \tag{5.5}$$

and reduces the problem to the calculation of the nth moment of the function of the random variable Y in the neighborhood of $n = 0$. Note that the expectation must be calculated for real-valued variables n in order to perform the limit operation.

At this point, it is customary to assume analytic continuity of the function $\mathrm{E}_Y[f(Y)]^n$. That is, the expectation is calculated for integer n only, but the resulting formula is trusted to hold for arbitrary real variables n in the neighborhood of $n = 0$. Note that analytic continuity is just an assumption. There is no mathematical theorem which states under which exact conditions this assumption is true or false. In fact, establishing a rigorous mathematical fundament for this step in the replica analysis is a topic of ongoing research. *However, in all physical problems where replicas have been introduced this procedure seems to work and leads to reasonable solutions* [5].

Relying on the analytic continuity, let the function

$$f(Y) = \int \mathrm{e}^{-||x||}\,\mathrm{d}x \tag{5.6}$$

take the form of a partition function where the dependency on Y is implicit via the definition of the energy function $||\cdot||$. Since the variable of integration is arbitrary, this implies

$$\mathop{\mathrm{E}}_Y [f(Y)]^n = \mathop{\mathrm{E}}_Y \left(\int \mathrm{e}^{-||x||}\,\mathrm{d}x \right)^n \tag{5.7}$$

$$= \mathop{\mathrm{E}}_Y \prod_{a=1}^{n} \int \mathrm{e}^{-||x_a||}\,\mathrm{d}x_a \tag{5.8}$$

$$= \int \mathop{\mathrm{E}}_Y \mathrm{e}^{-\sum\limits_{a=1}^{n} ||x_a||} \prod_{a=1}^{n} \mathrm{d}x_a. \tag{5.9}$$

Thus, instead of calculating the nth power of $f(Y)$, replicas of x are generated. These replicated variables x_a are arbitrary and can be assigned helpful properties. Often they are assumed to be independent random variables. Moreover, the replica trick allowed us to interchange integration and expectation, although the expectation in (5.7) is to be taken over a nonlinear function of the integral.

6. Saddle Point Integration

Typically, integrals arising from the replica ansatz are solved by saddle point integration. The general idea of saddle point integration is as follows: Consider an integral of the form

$$\frac{1}{K} \log \int \mathrm{e}^{Kf(x)}\mathrm{d}x. \tag{6.1}$$

Note that we can write this integral as

$$\frac{1}{K}\log\int e^{Kf(x)}\mathrm{d}x = \log\left[\int \exp(f(x))^K \mathrm{d}x\right]^{1/K} \tag{6.2}$$

which shows that it is actually the logarithm of the K-norm of the function $\exp(f(x))$. For $K \to \infty$, we get the maximum norm and thus obtain

$$\lim_{K\to\infty}\frac{1}{K}\log\int e^{Kf(x)}\mathrm{d}x = \max_x f(x). \tag{6.3}$$

That means, the integral can be solved maximizing the argument of the exponential function.

Some authors also refer to the saddle point integration as saddle point *approximation* and motivate it by a series expansion of the function $f(x)$ in the exponent. Making use of the identity (5.5) instead of (5.4), we can argue via the infinity norm and need not study under which conditions of the function $f(x)$ the saddle point approximation is accurate.

7. Replica Symmetry

If the function in the exponent is multivariate — typically all replicated random variables are arguments — one would need to find the extremum of a multivariate function for an arbitrary number of arguments. This can easily become a hopeless task, unless one can exploit some symmetries of the optimization problem.

Assuming *replica symmetry* means that one concludes from the symmetry of the exponent, e.g. $f(x_1, x_2) = f(x_2, x_1)$ for the bi-variate case, that the extremum appears if all variables take on the same value. Then, the multivariate optimization problem reduces to a single variate one, e.g.

$$\max_{x_1,x_1} f(x_1, x_2) = \max_x f(x, x) \tag{7.1}$$

for the bi-variate case. This is the most critical assumption when applying the replica method. In fact, it is not always true, even in practically relevant cases. Figure 1 shows both an example and a counterexample. The general way to circumvent this trouble is to assume replica symmetry at hand and proof later, having found a replica symmetric solution, that it is correct.

With the example of Fig. 1 in mind, it might seem that replica symmetry is a very odd assumption. However, the functions to be extremized arise from replication of identical integrals, see (5.9). Given the particular

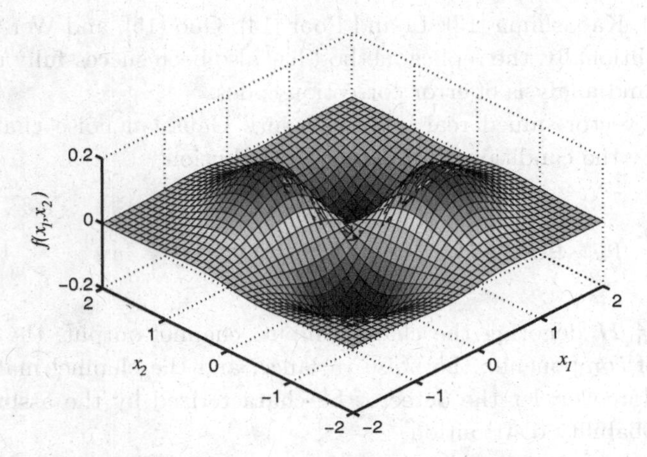

Fig. 1. Graph of the function $f(x_1, x_2) = -\sin(x_1 x_2)\exp(-x_1^2 - x_2^2)$. It is totally symmetric with respect to the exchange of x_1 and x_2. It shows symmetry with respect to its minima, but breaks symmetry with respect to its maxima.

structure of the optimization problem

$$\max_{x_1,\ldots,x_n} \prod_{a=1}^{n} f(x_i) \tag{7.2}$$

it seems rather odd that replica symmetry might *not* hold. However, writing our problem in the form of (7.2) assumes that the parameter n is an integer despite the fact that it is actually a real number in the neighborhood of zero. Thus, our intuition suggesting not to question replica symmetry cheats on us. In fact, there are even practically relevant cases without sensible replica symmetric solutions, e.g. cases were the replica symmetric solution implies the entropy to be negative. Such phenomena are labeled *replica symmetry breaking* and a rich theory in statistical mechanics literature exists to deal with them [6, 5, 1]. For the introductory character of this work, however, replica symmetry breaking is a too advanced issue.

8. Example: Analysis of Large CDMA Systems

The replica method was introduced into multiuser communications by the landmark paper of Tanaka [7] for the purpose of studying the performance of the maximum a-posteriori detector. Subsequently his work was generalized and extended other problems in multiuser communications by himself and Saad [8], Guo and Verdú [9], Müller *et al.* [10, 11], Caire *et al.* [4], Tanaka and Okada [12], Kabashima [13], Li and Poor [14], Guo [15], and Wen and Wong [16]. Additionally, the replica method has also been successfully used for the design and analysis of error correction codes.

Consider a vector-valued real additive white Gaussian noise channel characterized by the conditional probability distribution[b]

$$\mathrm{p}_{\boldsymbol{y}|\boldsymbol{x},\boldsymbol{H}}(\boldsymbol{y},\boldsymbol{x},\boldsymbol{H}) = \frac{\mathrm{e}^{-\frac{1}{2\sigma_0^2}(\boldsymbol{y}-\boldsymbol{H}\boldsymbol{x})^{\mathrm{T}}(\boldsymbol{y}-\boldsymbol{H}\boldsymbol{x})}}{(2\pi\sigma_0^2)^{\frac{N}{2}}} \tag{8.1}$$

with $\boldsymbol{x}, \boldsymbol{y}, N, \sigma_0^2, \boldsymbol{H}$ denoting the channel input, channel output, the latter's number of components, the noise variance, and the channel matrix, respectively. Moreover, let the detector be characterized by the assumed conditional probability distribution

$$\breve{\mathrm{p}}_{\boldsymbol{y}|\boldsymbol{x},\boldsymbol{H}}(\boldsymbol{y},\boldsymbol{x},\boldsymbol{H}) = \frac{\mathrm{e}^{-\frac{1}{2\sigma^2}(\boldsymbol{y}-\boldsymbol{H}\boldsymbol{x})^{\mathrm{T}}(\boldsymbol{y}-\boldsymbol{H}\boldsymbol{x})}}{(2\pi\sigma^2)^{\frac{N}{2}}} \tag{8.2}$$

and the assumed prior distribution $\breve{\mathrm{p}}_{\boldsymbol{x}}(\boldsymbol{x})$. Let the entries of $\boldsymbol{H}$ be independent zero-mean with vanishing odd order moments and variances w_{ck}/N for row c and column k. Moreover, let w_{ck} be uniformly bounded from above. Applying Bayes' law, we find

$$\breve{\mathrm{p}}_{\boldsymbol{x}|\boldsymbol{y},\boldsymbol{H}}(\boldsymbol{x},\boldsymbol{y},\boldsymbol{H}) = \frac{\mathrm{e}^{-\frac{1}{2\sigma^2}(\boldsymbol{y}-\boldsymbol{H}\boldsymbol{x})^{\mathrm{T}}(\boldsymbol{y}-\boldsymbol{H}\boldsymbol{x})+\log\breve{\mathrm{p}}_{\boldsymbol{x}}(\boldsymbol{x})}}{\int \mathrm{e}^{-\frac{1}{2\sigma^2}(\boldsymbol{y}-\boldsymbol{H}\boldsymbol{x})^{\mathrm{T}}(\boldsymbol{y}-\boldsymbol{H}\boldsymbol{x})}\,\mathrm{d}\breve{\mathrm{P}}_{\boldsymbol{x}}(\boldsymbol{x})}. \tag{8.3}$$

Since (3.3) holds for any temperature T, we set without loss of generality $T = 1$ in (3.3) and find the appropriate energy function to be

$$||\boldsymbol{x}|| = \frac{1}{2\sigma^2}(\boldsymbol{y}-\boldsymbol{H}\boldsymbol{x})^{\mathrm{T}}(\boldsymbol{y}-\boldsymbol{H}\boldsymbol{x}) - \log\breve{\mathrm{p}}_{\boldsymbol{x}}(\boldsymbol{x}). \tag{8.4}$$

This choice of the energy function ensures that the thermodynamic equilibrium models the detector defined by the assumed conditional and prior distributions.

Let K denote the number of users, that is the dimensionality of the input vector $\boldsymbol{x}$. Applying successively (4.3) with (8.1), (5.5), replica continuity

[b] In this example, we do not use upper case and lower case notation to distinguish random variables and their realizations to not mix up vectors and matrices.

(5.9), and (8.4) we find for the free energy per user

$$\frac{\mathrm{F}(\boldsymbol{x})}{K} = -\frac{1}{K} \mathop{\mathrm{E}}_{\boldsymbol{H}} \iint_{\mathbb{R}^N} \frac{\mathrm{e}^{-\frac{1}{2\sigma_0^2}(\boldsymbol{y}-\boldsymbol{H}\boldsymbol{x})^{\mathrm{T}}(\boldsymbol{y}-\boldsymbol{H}\boldsymbol{x})}}{(2\pi\sigma_0^2)^{\frac{N}{2}}} \log \int_{\mathbb{R}^K} \mathrm{e}^{-||\boldsymbol{x}||} \mathrm{d}\boldsymbol{x} \mathrm{d}\boldsymbol{y} \mathrm{dP}_{\boldsymbol{x}}(\boldsymbol{x})$$

$$= -\frac{1}{K} \lim_{n\to 0} \frac{\partial}{\partial n} \log \mathop{\mathrm{E}}_{\boldsymbol{H}} \iint_{\mathbb{R}^N} \left(\int_{\mathbb{R}^K} \mathrm{e}^{-||\boldsymbol{x}||} \mathrm{d}\boldsymbol{x} \right)^n$$

$$\times \frac{\mathrm{e}^{-\frac{1}{2\sigma_0^2}(\boldsymbol{y}-\boldsymbol{H}\boldsymbol{x})^{\mathrm{T}}(\boldsymbol{y}-\boldsymbol{H}\boldsymbol{x})}}{(2\pi\sigma_0^2)^{\frac{N}{2}}} \mathrm{d}\boldsymbol{y} \mathrm{dP}_{\boldsymbol{x}}(\boldsymbol{x}) \tag{8.5}$$

$$= -\frac{1}{K} \lim_{n\to 0} \frac{\partial}{\partial n} \log \underbrace{\frac{\int_{\mathbb{R}^N} \mathop{\mathrm{E}}_{\boldsymbol{H}} \prod_{a=0}^{n} \int \mathrm{e}^{-\frac{1}{2\sigma_a^2}(\boldsymbol{y}-\boldsymbol{H}\boldsymbol{x}_a)^{\mathrm{T}}(\boldsymbol{y}-\boldsymbol{H}\boldsymbol{x}_a)} \mathrm{dP}_a(\boldsymbol{x}_a)\, \mathrm{d}\boldsymbol{y}}{(2\pi\sigma_0^2)^{\frac{N}{2}}}}_{\triangleq \Xi_n}$$

with $\sigma_a = \sigma, \forall a \geq 1$, $\mathrm{P}_0(\boldsymbol{x}) = \mathrm{P}_{\boldsymbol{x}}(\boldsymbol{x})$, and $\mathrm{P}_a(\boldsymbol{x}) = \breve{\mathrm{P}}_{\boldsymbol{x}}(\boldsymbol{x}), \forall a \geq 1$.

The following calculations are a generalization of the derivations by Tanaka [7], Caire *et al.* [4], and Guo and Verdú [9]. They can also be found in a recent work of Guo [15]. The integral in (8.5) is given by

$$\Xi_n = \prod_{c=1}^{N} \frac{\int_{\mathbb{R}} \mathop{\mathrm{E}}_{\boldsymbol{H}} \prod_{a=0}^{n} \int \exp\left[-\frac{1}{2\sigma_a^2} \left(y_c - \sum_{k=1}^{K} h_{ck} x_{ak} \right)^2 \right] \mathrm{dP}_a(\boldsymbol{x}_a)\, \mathrm{d}y_c}{\sqrt{2\pi}\sigma_0}, \tag{8.6}$$

with y_c, x_{ak}, and h_{ck} denoting the cth component of $\boldsymbol{y}$, the kth component of $\boldsymbol{x}_a$, and the (c,k)th entry of $\boldsymbol{H}$, respectively. The integrand depends on $\boldsymbol{x}_a$ only through

$$v_{ac} \triangleq \frac{1}{\sqrt{\beta}} \sum_{k=1}^{K} h_{ck} x_{ak}, \qquad a = 0, \ldots, n \tag{8.7}$$

with the load β being defined as $\beta \triangleq K/N$. Following [7], the quantities v_{ac} can be regarded, in the limit $K \to \infty$ as jointly Gaussian random variables with zero mean and covariances

$$Q_{ab}[c] = \mathop{\mathrm{E}}_{\boldsymbol{H}} v_{ac} v_{bc} = \frac{1}{K} \boldsymbol{x}_a \overset{(c)}{\bullet} \boldsymbol{x}_b \tag{8.8}$$

where we defined the following inner products

$$\boldsymbol{x}_a \overset{(c)}{\bullet} \boldsymbol{x}_b \triangleq \sum_{k=1}^{K} x_{ak} x_{bk} w_{ck}. \tag{8.9}$$

In order to perform the integration in (8.6), the $K(n+1)$-dimensional space spanned by the replicas and the vector $\boldsymbol{x}_0$ is split into subshells

$$S\{Q[\cdot]\} \triangleq \left\{ \boldsymbol{x}_0, \ldots, \boldsymbol{x}_n \,\middle|\, \boldsymbol{x}_a \overset{(c)}{\bullet} \boldsymbol{x}_b = K Q_{ab}[c] \right\} \tag{8.10}$$

where the inner product of two different vectors $\boldsymbol{x}_a$ and $\boldsymbol{x}_b$ is constant.[c] The splitting of the $K(n+1)$-dimensional space is depending on the chip time c. With this splitting of the space, we find[d]

$$\Xi_n = \int_{\mathbb{R}^{N(n+1)(n+2)/2}} e^{K\mathcal{I}\{Q[\cdot]\}} \prod_{c=1}^{N} e^{\mathcal{G}\{Q[c]\}} \prod_{a \le b} dQ_{ab}[c], \tag{8.11}$$

where

$$e^{K\mathcal{I}\{Q[\cdot]\}} = \int \left[\prod_{a \le b} \prod_{c=1}^{N} \delta \left(\frac{\boldsymbol{x}_a \overset{(c)}{\bullet} \boldsymbol{x}_b}{N} - \beta Q_{ab}[c] \right) \right] \prod_{a=0}^{n} d\mathrm{P}_a(\boldsymbol{x}_a) \tag{8.12}$$

denotes the probability weight of the subshell and

$$e^{\mathcal{G}\{Q[c]\}} = \frac{1}{\sqrt{2\pi}\sigma_0} \int_{\mathbb{R}} \mathop{\mathrm{E}}_{\boldsymbol{H}} \prod_{a=0}^{n} \exp \left[-\frac{\beta}{2\sigma_a^2} \left(\frac{y_c}{\sqrt{\beta}} - v_{ac}\{Q[c]\} \right)^2 \right] dy_c. \tag{8.13}$$

This procedure is a change of integration variables in multiple dimensions where the integration of an exponential function over the replicas has been replaced by integration over the variables $\{Q[\cdot]\}$. In the following the two exponential terms in (8.11) are evaluated separately.

First, we turn to the evaluation of the measure $e^{K\mathcal{I}\{Q[\cdot]\}}$. Since for some $t \in \mathbb{R}$, we have the Fourier expansions of the Dirac measure

$$\delta \left(\frac{\boldsymbol{x}_a \overset{(c)}{\bullet} \boldsymbol{x}_b}{N} - \beta Q_{ab}[c] \right) = \frac{1}{2\pi j} \int_{\mathcal{J}} \exp \left[\tilde{Q}_{ab}[c] \left(\frac{\boldsymbol{x}_a \overset{(c)}{\bullet} \boldsymbol{x}_b}{N} - \beta Q_{ab}[c] \right) \right] d\tilde{Q}_{ab}[c] \tag{8.14}$$

[c]The notation $f\{Q[\cdot]\}$ expresses the dependency of the function $f(\cdot)$ on all $Q_{ab}[c], 0 \le a \le b \le n, 1 \le c \le N$.

[d]The notation $\prod_{a \le b}$ is used as shortcut for $\prod_{a=0}^{n} \prod_{b=a}^{n}$.

with $\mathcal{J} = (t - \mathrm{j}\infty; t + \mathrm{j}\infty)$, the measure $\mathrm{e}^{K\mathcal{I}\{Q[\cdot]\}}$ can be expressed as

$$\mathrm{e}^{K\mathcal{I}\{Q[\cdot]\}} = \int \left[\prod_{c=1}^{N} \prod_{a \le b} \int_{\mathcal{J}} \mathrm{e}^{\tilde{Q}_{ab}[c]\left(\frac{\boldsymbol{x}_a \overset{(c)}{\bullet} \boldsymbol{x}_b}{N} - \beta Q_{ab}[c] \right)} \frac{\mathrm{d}\tilde{Q}_{ab}[c]}{2\pi \mathrm{j}} \right] \prod_{a=0}^{n} \mathrm{dP}_a(\boldsymbol{x}_a) \tag{8.15}$$

$$= \int_{\mathcal{J}^{N(n+2)(n+1)/2}} \mathrm{e}^{-\beta \sum\limits_{c=1}^{N} \sum\limits_{a \le b} \tilde{Q}_{ab}[c] Q_{ab}[c]}$$

$$\times \left(\prod_{k=1}^{K} M_k \left\{ \tilde{Q}[\cdot] \right\} \right) \prod_{c=1}^{N} \prod_{a \le b} \frac{\mathrm{d}\tilde{Q}_{ab}[c]}{2\pi \mathrm{j}} \tag{8.16}$$

with

$$M_k \left\{ \tilde{Q}[\cdot] \right\} = \int \exp \left(\frac{1}{N} \sum_{a \le b} \sum_{c=1}^{N} \tilde{Q}_{ab}[c] x_{ak} x_{bk} w_{ck} \right) \prod_{a=0}^{n} \mathrm{dP}_a(x_{ak}). \tag{8.17}$$

In the limit of $K \to \infty$ one of the exponential terms in (8.11) will dominate over all others. Thus, only the maximum value of the correlation $Q_{ab}[c]$ is relevant for calculation of the integral, as shown in Section 6.

At this point, we assume that the replicas have a certain symmetry, as outlined in Section 7. This means, that in order to find the maximum of the objective function, we consider only a subset of the potential possibilities that the variables $Q_{ab}[\cdot]$ could take. Here, we restrict them to the following four different possibilities $Q_{00}[c] = p_{0c}$, $Q_{0a}[c] = m_c, \forall a \neq 0$, $Q_{aa}[c] = p_c, \forall a \neq 0$, $Q_{ab}[c] = q_c, \forall 0 \neq a \neq b \neq 0$. One case distinction has been made, as zero and non-zero replica indices correspond to the true and the assumed distributions, respectively, and thus will differ, in general. Another case distinction has been made to distinguish correlations $Q_{ab}[\cdot]$ which correspond to correlations between different and identical replica indices. This gives four cases to consider in total. We apply the same idea to the correlation variables in the Fourier domain and set $\tilde{Q}_{00}[c] = G_{0c}/2$, $\tilde{Q}_{aa}[c] = G_c/2, \forall a \neq 0$, $\tilde{Q}_{0a}[c] = E_c, \forall a \neq 0$, and $\tilde{Q}_{ab}[c] = F_c, \forall 0 \neq a \neq b \neq 0$.

At this point the crucial benefit of the replica method becomes obvious. Assuming replica continuity, we have managed to reduce the evaluation of a continuous function to sampling it at integer points. Assuming replica symmetry we have reduced the task of evaluating infinitely many integer points to calculating eight different correlations (four in the original and four in the Fourier domain).

The assumption of replica symmetry leads to

$$\sum_{a\leq b} \tilde{Q}_{ab}[c]Q_{ab}[c] = nE_c m_c + \frac{n(n-1)}{2}F_c q_c + \frac{G_{0c}p_{0c}}{2} + \frac{n}{2}G_c p_c \tag{8.18}$$

and

$$\begin{aligned} M_k\{E,F,G,G_0\} \\ &= \int \mathrm{e}^{\frac{1}{N}\sum\limits_{c=1}^{N} w_{ck}\left(\frac{G_{0c}}{2}x_{0k}^2 + \sum\limits_{a=1}^{n} E_c x_{0k}x_{ak} + \frac{G_c}{2}x_{ak}^2 + \sum\limits_{b=a+1}^{n} F_c x_{ak}x_{bk}\right)} \prod_{a=0}^{n} \mathrm{dP}_a(x_{ak}) \\ &= \int \mathrm{e}^{\frac{\tilde{G}_{0k}}{2}x_{0k}^2 + \sum\limits_{a=1}^{n} \tilde{E}_k x_{0k}x_{ak} + \frac{\tilde{G}_k}{2}x_{ak}^2 + \sum\limits_{b=a+1}^{n} \tilde{F}_k x_{ak}x_{bk}} \prod_{a=0}^{n} \mathrm{dP}_a(x_{ak}) \end{aligned} \tag{8.19}$$

where

$$\tilde{E}_k \triangleq \frac{1}{N}\sum_{c=1}^{N} E_c w_{ck} \tag{8.20}$$

$$\tilde{F}_k \triangleq \frac{1}{N}\sum_{c=1}^{N} F_c w_{ck} \tag{8.21}$$

$$\tilde{G}_k \triangleq \frac{1}{N}\sum_{c=1}^{N} G_c w_{ck} \tag{8.22}$$

$$\tilde{G}_{0k} \triangleq \frac{1}{N}\sum_{c=1}^{N} G_{0c} w_{ck}. \tag{8.23}$$

Note that the prior distribution enters the free energy only via the (8.19). We will focus on this later on after having finished with the other terms.

For the evaluation of $\mathrm{e}^{\mathcal{G}\{Q[c]\}}$ in (8.11), we can use the replica symmetry to construct the correlated Gaussian random variables v_{ac} out of independent zero-mean, unit-variance Gaussian random variables u_c, t_c, z_{ac} by

$$v_{0c} = u_c\sqrt{p_{0c} - \frac{m_c^2}{q_c}} - t_c\frac{m_c}{\sqrt{q_c}} \tag{8.24}$$

$$v_{ac} = z_{ac}\sqrt{p_c - q_c} - t_c\sqrt{q_c}, \quad a > 0. \tag{8.25}$$

With that substitution, we get

$$e^{\mathcal{G}(m_c,q_c,p_c,p_{0c})} \tag{8.26}$$

$$= \frac{1}{\sqrt{2\pi}\sigma_0}\int_{\mathbb{R}^2}\int_{\mathbb{R}} \exp\left[-\frac{\beta}{2\sigma_0^2}\left(u_c\sqrt{p_{0c}-\frac{m_c^2}{q_c}}-\frac{t_c m_c}{\sqrt{q_c}}-\frac{y_c}{\sqrt{\beta}}\right)^2\right] \mathrm{D}u_c$$

$$\times\left[\int_{\mathbb{R}}\exp\left[-\frac{\beta}{2\sigma^2}\left(z_c\sqrt{p_c-q_c}-t_c\sqrt{q_c}-\frac{y_c}{\sqrt{\beta}}\right)^2\right]\mathrm{D}z_c\right]^n \mathrm{D}t_c\,\mathrm{d}y_c$$

$$= \sqrt{\frac{\left(1+\frac{\beta}{\sigma^2}(p_c-q_c)\right)^{1-n}}{1+\frac{\beta}{\sigma^2}(p_c-q_c)+n\frac{\beta}{\sigma^2}\left(\frac{\sigma_0^2}{\beta}+p_{0c}-2m_c+q_c\right)}} \tag{8.27}$$

with the Gaussian measure $\mathrm{D}z = \exp(-z^2/2)\,\mathrm{d}z/\sqrt{2\pi}$. Since the integral in (8.11) is dominated by the maximum argument of the exponential function, the derivatives of

$$\frac{1}{N}\sum_{c=1}^{N}\left(\mathcal{G}\{Q[c]\}-\beta\sum_{a\le b}\tilde{Q}_{ab}[c]Q_{ab}[c]\right) \tag{8.28}$$

with respect to m_c, q_c, p_c and p_{0c} must vanish as $N \to \infty$. Taking derivatives after plugging (8.18) and (8.27) into (8.28), solving for E_c, F_c, G_c, and G_{0c} and letting $n \to 0$ yields for all c

$$E_c = \frac{1}{\sigma^2+\beta(p_c-q_c)} \tag{8.29}$$

$$F_c = \frac{\sigma_0^2+\beta\,(p_{0c}-2m_c+q_c)}{[\sigma^2+\beta(p_c-q_c)]^2} \tag{8.30}$$

$$G_c = F_c - E_c \tag{8.31}$$

$$G_{0c} = 0. \tag{8.32}$$

In the following, the calculations are shown explicitly for Gaussian and binary priors. Additionally, a general formula for arbitrary priors is given.

8.1. *Gaussian prior distribution*

Assume a Gaussian prior distribution

$$\mathrm{p}_a(x_{ak}) = \frac{1}{\sqrt{2\pi}}\,\mathrm{e}^{-x_{ak}^2/2} \qquad \forall a. \tag{8.33}$$

Thus, the integration in (8.19) can be performed explicitly and we find with [7, (87)]

$$M_k\left\{E,F,G,G_0\right\}=\sqrt{\frac{\left(1+\tilde{F}_k-\tilde{G}_k\right)^{1-n}}{\left(1-\tilde{G}_{0k}\right)\left(1+\tilde{F}_k-\tilde{G}_k-n\tilde{F}_k\right)-n\tilde{E}_k^2}}. \tag{8.34}$$

In the large system limit, the integral in (8.16) is also dominated by that value of the integration variable which maximizes the argument of the exponential function under some weak conditions on the variances w_{ck}. Thus, partial derivatives of

$$\log\prod_{k=1}^{K}M_k\{E,F,G,G_0\}-\beta\sum_{c=1}^{N}nE_cm_c+\frac{n(n-1)}{2}F_cq_c+\frac{G_{0c}p_{0c}}{2}+\frac{n}{2}G_cp_c \tag{8.35}$$

with respect to E_c, F_c, G_c, G_{0c} must vanish for all c as $N\to\infty$. An explicit calculation of these derivatives yields

$$m_c=\frac{1}{K}\sum_{k=1}^{K}w_{ck}\,\frac{\tilde{E}_k}{1+\tilde{E}_k} \tag{8.36}$$

$$q_c=\frac{1}{K}\sum_{k=1}^{K}w_{ck}\,\frac{\tilde{E}_k^2+\tilde{F}_k}{\left(1+\tilde{E}_k\right)^2} \tag{8.37}$$

$$p_c=\frac{1}{K}\sum_{k=1}^{K}w_{ck}\,\frac{\tilde{E}_k^2+\tilde{E}_k+\tilde{F}_k+1}{\left(1+\tilde{E}_k\right)^2} \tag{8.38}$$

$$p_{0c}=\frac{1}{K}\sum_{k=1}^{K}w_{ck} \tag{8.39}$$

in the limit $n\to 0$ with (8.31) and (8.32). Surprisingly, if we let the true prior to be binary and only the replicas to be Gaussian we also find (8.36) to (8.39). This setting corresponds to linear MMSE detection [17].

Returning to our initial goal, the evaluation of the free energy, and collecting our previous results, we find

$$\frac{\mathrm{F}(\boldsymbol{x})}{K} = -\frac{1}{K}\lim_{n\to 0}\frac{\partial}{\partial n}\log \Xi_n \tag{8.40}$$

$$= \frac{1}{K}\lim_{n\to 0}\frac{\partial}{\partial n}\sum_{c=1}^{N}\left[-\mathcal{G}(m_c,q_c,p_c,p_{0c}) + \frac{\beta n(n-1)}{2}F_c q_c \right.$$
$$\left. + \beta n E_c m_c + \frac{\beta n}{2}G_c p_c\right] - \sum_{k=1}^{K}\log M_k\left\{E,F,G,0\right\} \tag{8.41}$$

$$= \frac{1}{2K}\lim_{n\to 0}\left[\sum_{c=1}^{N}\log\left(1+\frac{\beta}{\sigma^2}(p_c-q_c)\right) + 2\beta E_c m_c + \beta G_c p_c\right.$$
$$\left. + \beta(2n-1)F_c q_c + \frac{\sigma_0^2 + \beta(p_{0c}-2m_c+q_c)}{\sigma^2+\beta(p_c-q_c)+n\sigma_0^2+n\beta(p_{0c}-2m_c+q_c)}\right]$$
$$+\frac{1}{2K}\lim_{n\to 0}\sum_{k=1}^{K}\log\left(1+\tilde{E}_k\right) - \frac{\tilde{E}_k^2+\tilde{F}_k}{1+\tilde{E}_k-n\tilde{E}_k^2-n\tilde{F}_k} \tag{8.42}$$

$$= \frac{1}{2K}\left[\sum_{c=1}^{N}\log\left(1+\frac{\beta}{\sigma^2}(p_c-q_c)\right) + 2\beta E_c m_c - \beta F_c q_c\right.$$
$$\left. + \beta G_c p_c + \frac{E_c}{F_c}\right] + \frac{1}{2K}\sum_{k=1}^{K}\log\left(1+\tilde{E}_k\right) - \frac{\tilde{E}_k^2+\tilde{F}_k}{1+\tilde{E}_k}. \tag{8.43}$$

This is the final result for the free energy of the mismatched detector assuming noise variance σ^2 instead of the true noise variance σ_0^2. The six macroscopic parameters $E_c, F_c, G_c, m_c, q_c, p_c$ are implicitly given by the simultaneous solution of the system of equations (8.29) to (8.31) and (8.36) to (8.38) with the definitions (8.20) to (8.22) for all chip times c. This system of equations can only be solved numerically.

Specializing our result to the matched detector assuming the true noise variance by letting $\sigma \to \sigma_0$, we have $F_c \to E_c$, $G_c \to G_{0c}$, $q_c \to m_c$, $p_c \to p_{0c}$. This makes the free energy simplify to

$$\frac{\mathrm{F}(\boldsymbol{x})}{K} = \frac{1}{2K}\left[\sum_{c=1}^{N}\sigma_0^2 E_c - \log\left(\sigma_0^2 E_c\right)\right] + \frac{1}{2K}\sum_{k=1}^{K}\log\left(1+\tilde{E}_k\right) \tag{8.44}$$

with

$$E_c = \frac{1}{\sigma_0^2 + \frac{\beta}{K}\sum_{k=1}^{K}\frac{w_{ck}}{1+\tilde{E}_k}}. \tag{8.45}$$

This result is more compact and it requires only to solve (8.45) numerically which is conveniently done by fixed-point iteration.

It can be shown that the parameter $\tilde{E}_k$ is actually the signal-to-interference and noise ratio of user k. It has been derived independently by Hanly and Tse [18] in context of CDMA with macro-diversity using a result from random matrix theory by Girko [19]. Note that (8.45) and (8.20) are actually formally equivalent to the result Girko found for random matrices.

The similarity of free energy with the entropy of the channel output mentioned at the end of Section 4 is expressed by the simple relationship

$$\frac{\mathrm{I}(\boldsymbol{x},\boldsymbol{y})}{K} = \frac{\mathrm{F}(\boldsymbol{x})}{K} - \frac{1}{2\beta} \tag{8.46}$$

between the (normalized) free energy and the (normalized) mutual information between channel input signal $\boldsymbol{x}$ and channel output signal $\boldsymbol{y}$ given the channel matrix $\boldsymbol{H}$. Assuming that the channel is perfectly known to the receiver, but totally unknown to the transmitter, (8.46) gives the channel capacity per user.

8.2. *Binary prior distribution*

Now, we assume a non-uniform binary prior distribution

$$\mathrm{p}_a(x_{ak}) = \frac{1+t_k}{2}\,\delta(x_{ak}-1) + \frac{1-t_k}{2}\,\delta(x_{ak}+1). \tag{8.47}$$

Plugging the prior distribution into (8.19), we find

$$\begin{aligned}
&M_k\{E,F,G,G_0\} \\
&= \int_{\mathbb{R}^{n+1}} \mathrm{e}^{\frac{\tilde{G}_{0k}+n\tilde{G}_k}{2}+\sum\limits_{a=1}^{n}\tilde{E}_k x_{0k}x_{ak}+\sum\limits_{b=a+1}^{n}\tilde{F}_k x_{ak}x_{bk}} \prod_{a=0}^{n} \mathrm{dP}_a(x_{ak}) &(8.48)\\
&= \mathrm{e}^{\frac{1}{2}(\tilde{G}_{0k}+n\tilde{G}_k)} \sum_{\{x_{ak},a=1,\ldots,n\}} \left\{ \frac{1+t_k}{2}\exp\left[\sum_{a=1}^{n}\tilde{E}_k x_{ak} + \sum_{b=a+1}^{n}\tilde{F}_k x_{ak}x_{bk}\right]\right.\\
&\quad \left. + \frac{1-t_k}{2}\exp\left[\sum_{a=1}^{n} -\tilde{E}_k x_{ak} + \sum_{b=a+1}^{n}\tilde{F}_k x_{ak}x_{bk}\right]\right\} \prod_{a=1}^{n}\Pr(x_{ak}) &(8.49)\\
&= \mathrm{e}^{\frac{1}{2}(\tilde{G}_{0k}+n\tilde{G}_k-n\tilde{F}_k)} \sum_{\{x_{ak},a=1,\ldots,n\}} \left\{ \frac{1+t_k}{2}\exp\left[\frac{\tilde{F}_k}{2}\left(\sum_{a=1}^{n}x_{ak}\right)^2 + \tilde{E}_k\sum_{a=1}^{n}x_{ak}\right]\right.\\
&\quad \left. + \frac{1-t_k}{2}\exp\left[\frac{\tilde{F}_k}{2}\left(\sum_{a=1}^{n}x_{ak}\right)^2 - \tilde{E}_k\sum_{a=1}^{n}x_{ak}\right]\right\} \prod_{a=1}^{n}\Pr(x_{ak}) &(8.50)
\end{aligned}$$

where we can use the following property of the Gaussian measure

$$\exp\left(\tilde{F}_k \frac{S^2}{2}\right) = \int \exp\left(\pm\sqrt{\tilde{F}_k} zS\right) \mathrm{D}z \qquad \forall S \in \mathbb{R} \tag{8.51}$$

which is also called the Hubbard-Stratonovich transform to linearize the exponents

$$\begin{aligned} &M_k\left\{E,F,G,G_0\right\} \\ &= e^{\frac{1}{2}(\tilde{G}_{0k}+n\tilde{G}_k-n\tilde{F}_k)} \sum_{\{x_{ak},a=1,\ldots,n\}} \int \frac{1+t_k}{2} \exp\left[\left(z\sqrt{\tilde{F}_k}+\tilde{E}_k\right)\sum_{a=1}^n x_{ak}\right] \\ &\quad + \frac{1-t_k}{2} \exp\left[-\left(z\sqrt{\tilde{F}_k}+\tilde{E}_k\right)\sum_{a=1}^n x_{ak}\right] \mathrm{D}z \prod_{a=1}^n \Pr(x_{ak}). \end{aligned} \tag{8.52}$$

Since

$$\begin{aligned} f_n &\triangleq \sum_{\{x_{ka},a=1,\ldots,n\}} \exp\left[\left(z\sqrt{\tilde{F}_k}+\tilde{E}_k\right)\sum_{a=1}^n x_{ka}\right] \prod_{a=1}^n \Pr(x_{ka}) && (8.53) \\ &= \sum_{x_{kn}} \Pr(x_{kn}) f_{n-1} \exp\left[\left(z\sqrt{\tilde{F}_k}+\tilde{E}_k\right) x_{kn}\right] && (8.54) \\ &= f_{n-1} \frac{\cosh\left(\lambda_k/2 + z\sqrt{\tilde{F}_k}+\tilde{E}_k\right)}{\cosh\left(\lambda_k/2\right)} && (8.55) \\ &= \frac{\cosh^n\left(\lambda_k/2 + z\sqrt{\tilde{F}_k}+\tilde{E}_k\right)}{\cosh^n\left(\lambda_k/2\right)} && (8.56) \end{aligned}$$

with $t_k \triangleq \tanh(\lambda_k/2)$, we find

$$\begin{aligned} &M_k\left\{E,F,G,G_0\right\} \\ &= \frac{\int \frac{1+t_k}{2} \cosh^n\left(z\sqrt{\tilde{F}_k}+\tilde{E}_k+\frac{\lambda_k}{2}\right) + \frac{1-t_k}{2} \cosh^n\left(z\sqrt{\tilde{F}_k}+\tilde{E}_k-\frac{\lambda_k}{2}\right) \mathrm{D}z}{\cosh^n\left(\frac{\lambda_k}{2}\right) \exp\left(\frac{n\tilde{F}_k-\tilde{G}_{0k}-n\tilde{G}_k}{2}\right)}. \end{aligned} \tag{8.57}$$

In the large system limit, the integral in (8.16) is dominated by that value of the integration variable which maximizes the argument of the exponential function under some weak conditions on the variances w_{ck}. Thus, partial derivations of (8.35) with respect to E_c, F_c, G_c, G_{0c} must vanish for all c as $N \to \infty$. An explicit calculation of these derivatives gives

$$m_c = \frac{1}{K}\sum_{k=1}^{K} w_{ck} \int \frac{1+t_k}{2} \tanh\left(z\sqrt{\tilde{F}_k} + \tilde{E}_k + \frac{\lambda_k}{2}\right) + \frac{1-t_k}{2} \tanh\left(z\sqrt{\tilde{F}_k} + \tilde{E}_k - \frac{\lambda_k}{2}\right) \mathrm{D}z \tag{8.58}$$

$$q_c = \frac{1}{K}\sum_{k=1}^{K} w_{ck} \int \frac{1+t_k}{2} \tanh^2\left(z\sqrt{\tilde{F}_k} + \tilde{E}_k + \frac{\lambda_k}{2}\right) + \frac{1-t_k}{2} \tanh^2\left(z\sqrt{\tilde{F}_k} + \tilde{E}_k - \frac{\lambda_k}{2}\right) \mathrm{D}z \tag{8.59}$$

$$p_c = p_{0c} = \frac{1}{K}\sum_{k=1}^{K} w_{ck} \tag{8.60}$$

in the limit $n \to 0$. In order to obtain (8.59), note from (8.50) that the first order derivative of $M_k \exp(n\tilde{F}_k/2)$ with respect to F_c is identical to half of the second order derivative of $M_k \exp(n\tilde{F}_k/2)$ with respect to E_c.

Returning to our initial goal, the evaluation of the free energy, and collecting our previous results, we find

$$\begin{aligned}
\frac{\mathrm{F}(\boldsymbol{x})}{K} &= -\frac{1}{K}\lim_{n\to 0}\frac{\partial}{\partial n}\log \Xi_n \\
&= \frac{1}{K}\lim_{n\to 0}\frac{\partial}{\partial n}\sum_{c=1}^{N}\Bigg[-\mathcal{G}(m_c, q_c, p_c, p_{0c}) + \beta n E_c m_c \\
&\quad + \frac{\beta n(n-1)}{2} F_c q_c + \frac{\beta n}{2} G_c p_c\Bigg] - \sum_{k=1}^{K}\log M_k\{E, F, G, 0\} \qquad (8.61) \\
&= \frac{1}{2K}\sum_{c=1}^{N}\Bigg[\log\left(1 + \frac{\beta}{\sigma^2}(p_c - q_c)\right) \\
&\quad + \beta E_c(2m_c + p_c) + \beta F_c(p_c - q_c) + \frac{E_c}{F_c}\Bigg] \\
&\quad - \frac{1}{K}\sum_{k=1}^{K}\int \frac{1+t_k}{2}\log\cosh\left(z\sqrt{\tilde{F}_k} + \tilde{E}_k + \frac{\lambda_k}{2}\right) - \frac{\tilde{E}_k}{2} \\
&\quad + \frac{1-t_k}{2}\log\cosh\left(z\sqrt{\tilde{F}_k} + \tilde{E}_k - \frac{\lambda_k}{2}\right)\mathrm{D}z + \frac{1}{2}\log\left(1 - t_k^2\right).
\end{aligned} \tag{8.62}$$

This is the final result for the free energy of the mismatched detector assuming noise variance σ^2 instead of the true noise variance σ_0^2. The five macroscopic parameters E_c, F_c, m_c, q_c, p_c are implicitly given by the simultaneous solution of the system of equations (8.29), (8.30) and (8.58) to (8.60) with the definitions (8.20) to (8.22) for all chip times c. This system of equations can only be solved numerically. Moreover, it can have multiple solutions. In case of multiple solutions, the correct solution is that one which minimizes the free energy, since in the thermodynamic equilibrium the free energy is always minimized, cf. Section 3.

Specializing our result to the matched detector assuming the true noise variance by letting $\sigma \to \sigma_0$, we have $F_c \to E_c$, $G_c \to G_{0c}$, $q_c \to m_c$ which makes the free energy simplify to

$$\begin{aligned}\frac{\mathrm{F}(\boldsymbol{x})}{K} = & \frac{1}{2K}\sum_{c=1}^{N}\left[\sigma_0^2 E_c - \log\left(\sigma_0^2 E_c\right)\right] - \frac{1}{K}\sum_{k=1}^{K}\frac{1}{2}\log\left(1-t_k^2\right) - \tilde{E}_k \\ & + \int \frac{1+t_k}{2}\log\cosh\left(z\sqrt{\tilde{E}_k} + \tilde{E}_k + \frac{\lambda_k}{2}\right) \\ & + \frac{1-t_k}{2}\log\cosh\left(z\sqrt{\tilde{E}_k} + \tilde{E}_k - \frac{\lambda_k}{2}\right)\mathrm{D}z\end{aligned} \qquad (8.63)$$

where the macroscopic parameters E_c are given by

$$\frac{1}{E_c} = \sigma_0^2 + \frac{\beta}{K}\sum_{k=1}^{K} w_{ck}\left(1-t_k^2\right)\int \frac{1-\tanh\left(z\sqrt{\tilde{E}_k}+\tilde{E}_k\right)}{1-t_k^2\tanh^2\left(z\sqrt{\tilde{E}_k}+\tilde{E}_k\right)}\mathrm{D}z. \qquad (8.64)$$

Similar to the case of Gaussian priors, $\tilde{E}_k$ can be shown to be a kind of signal-to-interference and noise ratio, in the sense that the bit error probability of user k is given by

$$\Pr(\hat{x}_k \neq x_k) = \int_{\sqrt{\tilde{E}_k}}^{\infty} \mathrm{D}z. \qquad (8.65)$$

In fact, it can even be shown that in the large system limit, an equivalent additive white Gaussian noise channel can be defined to model the multiuser interference [10, 9]. Constraining the input alphabet of the channel to follow the non-uniform binary distribution (8.47) and assuming channel state information being available only at the transmitter, channel capacity is given by (8.46) with the free energy given in (8.63).

Large system results for binary prior (even for uniform binary prior) have not yet been able to be derived by means of rigorous mathematics despite intense effort to do so. Only for the case of vanishing noise variance a fully mathematically rigorous result was found by Tse and Verdú [20] which does not rely on the replica method.

8.3. *Arbitrary prior distribution*

Consider now an arbitrary prior distribution. As shown by Guo and Verdú [9], this still allows to reduce the multi-dimensional integration over all replicated random variables to a scalar integration over the prior distribution. Consider (8.19) giving the only term that involves the prior distribution and apply the Hubbard-Stratonovich transform (8.51)

$$M_k\{E, F, G, G_0\}$$

$$= \int \mathrm{e}^{\frac{\tilde{G}_{0k}}{2} x_{0k}^2 + \sum_{a=1}^{n} \tilde{E}_k x_{0k} x_{ak} + \frac{\tilde{G}_k}{2} x_{ak}^2 + \sum_{b=a+1}^{n} \tilde{F}_k x_{ak} x_{bk}} \prod_{a=0}^{n} \mathrm{dP}_a(x_{ak}) \tag{8.66}$$

$$= \int \mathrm{e}^{\frac{\tilde{G}_{0k}}{2} x_{0k}^2 + \frac{\tilde{F}_k}{2} \left(\sum_{a=1}^{n} x_{ak} \right)^2 + \sum_{a=1}^{n} \tilde{E}_k x_{0k} x_{ak} + \frac{\tilde{G}_k - \tilde{F}_k}{2} x_{ak}^2} \prod_{a=0}^{n} \mathrm{dP}_a(x_{ak}) \tag{8.67}$$

$$= \iint \mathrm{e}^{\frac{\tilde{G}_{0k}}{2} x_{0k}^2 + \sum_{a=1}^{n} \tilde{E}_k x_{0k} x_{ak} + \sqrt{\tilde{F}_k} z x_{ak} + \frac{\tilde{G}_k - \tilde{F}_k}{2} x_{ak}^2} \mathrm{D}z \prod_{a=0}^{n} \mathrm{dP}_a(x_{ak}) \tag{8.68}$$

$$= \int \mathrm{e}^{\frac{\tilde{G}_{0k}}{2} x_k^2} \int \left(\int \mathrm{e}^{\tilde{E}_k x_k \breve{x}_k + \sqrt{\tilde{F}_k} z \breve{x}_k + \frac{\tilde{G}_k - \tilde{F}_k}{2} \breve{x}_k^2} \mathrm{d}\breve{\mathrm{P}}_{\breve{x}_k}(\breve{x}_k) \right)^n \mathrm{D}z \mathrm{dP}_{x_k}(x_k) . \tag{8.69}$$

In the large system limit, the integral in (8.16) is dominated by that value of the integration variable which maximizes the argument of the exponential function under some weak conditions on the variances w_{ck}. Thus, partial derivations of (8.35) with respect to E_c, F_c, G_c, G_{0c} must vanish for all c as $N \to \infty$. While taking derivatives with respect to E_c, G_c and G_{0c} straightforwardly lead to suitable results, the derivative with respect to F_c requires a little trick: Note for the integrand I_k in (8.67), we have

$$\frac{\partial}{\partial F_c} I_k = \frac{1}{2x_{0k}^2} \frac{\partial^2}{\partial E_c^2} I_k - \frac{\partial}{\partial G_c} I_k . \tag{8.70}$$

With the help of (8.70), an explicit calculation of the four derivatives gives the following expressions for the macroscopic parameters m_c, q_c, p_c and p_{0c}

$$m_c = \frac{1}{K}\sum_{k=1}^{K} w_{ck} \iint x_k \frac{\int \breve{x}_k \mathrm{e}^{\tilde{E}_k\left(x_k\breve{x}_k - \frac{\breve{x}_k^2}{2}\right)+\sqrt{\tilde{F}_k} z\breve{x}_k} \mathrm{d}\breve{\mathrm{P}}_{\breve{x}_k}(\breve{x}_k)}{\int \mathrm{e}^{\tilde{E}_k\left(x_k\breve{x}_k - \frac{\breve{x}_k^2}{2}\right)+\sqrt{\tilde{F}_k} z\breve{x}_k} \mathrm{d}\breve{\mathrm{P}}_{\breve{x}_k}(\breve{x}_k)} \mathrm{D}z\mathrm{dP}_{x_k}(x_k) \tag{8.71}$$

$$q_c = \frac{1}{K}\sum_{k=1}^{K} w_{ck} \iint \left[\frac{\int \breve{x}_k \mathrm{e}^{\tilde{E}_k\left(x_k\breve{x}_k - \frac{\breve{x}_k^2}{2}\right)+\sqrt{\tilde{F}_k} z\breve{x}_k} \mathrm{d}\breve{\mathrm{P}}_{\breve{x}_k}(\breve{x}_k)}{\int \mathrm{e}^{\tilde{E}_k\left(x_k\breve{x}_k - \frac{\breve{x}_k^2}{2}\right)+\sqrt{\tilde{F}_k} z\breve{x}_k} \mathrm{d}\breve{\mathrm{P}}_{\breve{x}_k}(\breve{x}_k)}\right]^2 \mathrm{D}z\mathrm{dP}_{x_k}(x_k) \tag{8.72}$$

$$p_c = \frac{1}{K}\sum_{k=1}^{K} w_{ck} \iint \frac{\int \breve{x}_k^2 \mathrm{e}^{\tilde{E}_k\left(x_k\breve{x}_k - \frac{\breve{x}_k^2}{2}\right)+\sqrt{\tilde{F}_k} z\breve{x}_k} \mathrm{d}\breve{\mathrm{P}}_{\breve{x}_k}(\breve{x}_k)}{\int \mathrm{e}^{\tilde{E}_k\left(x_k\breve{x}_k - \frac{\breve{x}_k^2}{2}\right)+\sqrt{\tilde{F}_k} z\breve{x}_k} \mathrm{d}\breve{\mathrm{P}}_{\breve{x}_k}(\breve{x}_k)} \mathrm{D}z\mathrm{dP}_{x_k}(x_k) \tag{8.73}$$

$$p_{0c} = \frac{1}{K}\sum_{k=1}^{K} w_{ck} \int x_k^2 \mathrm{dP}_{x_k}(x_k) \tag{8.74}$$

with (8.31) and (8.32) in the limit $n \to 0$.

Returning to our initial goal, the evaluation of the free energy, and collecting our previous results, we find

$$\frac{\mathrm{F}(\boldsymbol{x})}{K} = \frac{1}{2K}\sum_{c=1}^{N}\left[\log\left(1+\frac{\beta}{\sigma^2}(p_c-q_c)\right)+\beta E_c(2m_c+p_c)+\beta F_c(p_c-q_c)+\frac{E_c}{F_c}\right]$$
$$-\frac{1}{K}\sum_{k=1}^{K}\iint \log \int \mathrm{e}^{\tilde{E}_k\left(x_k\breve{x}_k - \frac{\breve{x}_k^2}{2}\right)+\sqrt{\tilde{F}_k} z\breve{x}_k} \mathrm{d}\breve{\mathrm{P}}_{\breve{x}_k}(\breve{x}_k)\, \mathrm{D}z\mathrm{dP}_{x_k}(x_k)\,. \tag{8.75}$$

This is the final result for the free energy of the mismatched detector assuming noise variance σ^2 instead of the true noise variance σ_0^2. The five macroscopic parameters E_c, F_c, m_c, q_c, p_c are implicitly given by the simultaneous solution of the system of equations (8.29), (8.30) and (8.58) to (8.60) with the definitions (8.20) to (8.22) for all chip times c. This system of equations can only be solved numerically. Moreover, it can have multiple solutions. In case of multiple solutions, the correct solution is that one which minimizes the free energy, since in the thermodynamic equilibrium the free energy is always minimized, cf. Section 3.

9. Phase Transitions

In thermodynamics, the occurrence of phase transitions, i.e. melting ice becomes water, is a well-known phenomenon. In digital communications, however, such phenomena are less known, though they do occur. The similarity between thermodynamics and multiuser detection pointed out in Section 4, should be sufficient to convince the reader that phase transitions in digital communications do occur. Phase transitions in turbo decoding and detection of CDMA were found in [21] and [7], respectively.

The phase transitions in digital communications are similar to the hysteresis in ferro-magnetic materials. They occur if the equations determining the macroscopic parameters, e.g. E_c determined by (8.64), have multiple solutions. Then, it is the free energy to decide which of the solution corresponds to the thermodynamic equilibrium. If a system parameter, e.g. the load or the noise variance, changes, the free energy may shift its favor from one solution to another one. Since each solution corresponds to a different macroscopic property of the system, changing the valid solution means that a phase transition takes place.

In digital communications, a popular macroscopic property is the bit error probability. It is related to the macroscopic property $\tilde{E}_k$ in (8.64) by (8.65) for the case considered in Section 8. Numerical results are depicted in Fig. 2. The thick curve shows the bit error probability of the individually optimum detector as a function of the load. The thin curves show alternative solutions for the bit error probability corresponding to alternative solutions to the equations for the macroscopic variable $\tilde{E}_k$. Only for a certain interval of the load, approximately $1.73 \leq \beta \leq 3.56$ in Fig. 2, multiple solutions coexist. As expected, the bit error probability increases with the load. At a load of approximately $\beta = 1.986$ a phase transition occurs and lets the bit error probability jump. Unlike to ferromagnetic materials, there is no hysteresis effect for the bit error probability of the individually optimum detector, but only a phase transition. This is, as the external magnetic field corresponds to the channel output observed by the receiver. Unlike an external magnetic field, the channel output is a statistical variable and cannot be design to undergo certain trajectories.

In order to observe a hysteresis behavior, we can expand our scope to neural networks. Consider a Hopfield neural network [22] implementation of

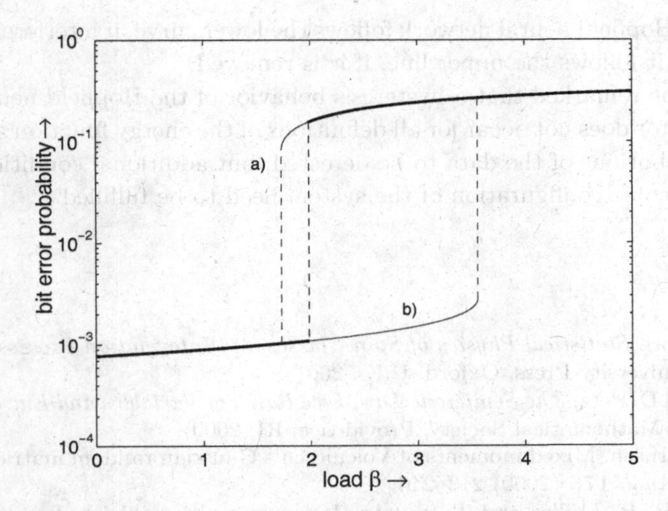

Fig. 2. Bit error probability for the individually optimum detector with uniform binary prior distribution versus system load for $10 \log_{10}(E_s/N_0) = 6$ dB.

the individually optimum multiuser detector which is an algorithm based on non-linear gradient search maximizing the energy function associated with the detector. Its application to the problem of multiuser detection is discussed in [23]. With appropriate definition of the energy function, such a detector will achieve the performance of the upper curve in Fig. 2 in the large system limit. Thus, in the interval $1.73 \leq \beta \leq 1.986$ where the free energy favors the curve with lower bit error probability, the Hopfield neural network is suboptimum (labeled with *a*)[e]. The curve labeled with *b* can also be achieved by the Hopfield neural network, but only with the help of a genie. In order to achieve a point in that area, cancel with the help of a genie as many interferers as needed to push the load below the area where multiple solutions occur, i.e. $\beta < 1.73$. Then, initialize the Hopfield neural network with the received vector where the interference has been canceled and let it converge to the thermodynamic equilibrium. Then, slowly add one by one the interferers you had canceled with the help of the genie while the Hopfield neural network remains in the thermodynamic equilibrium by performing iterations. If all the interference suppressed by the genie has been added again, the targeted point on the lower curve in area *b* is

[e]Note that in a system with a finite number of users, the Hopfield neural network is suboptimal at any load.

reached. The Hopfield neural network follows the lower curve, if interference is added, and it follows the upper line, if it is removed.

It should be remarked that a hysteresis behavior of the Hopfield neural network detector does not occur for all definitions of the energy function and all prior distributions of the data to be detected, but additional conditions on the microscopic configuration of the system need to be fulfilled.

References

1. H. Nishimori, *Statistical Physics of Spin Glasses and Information Processing* (Oxford University Press, Oxford, U.K., 2001).
2. F. Hiai and D. Petz, *The Semicircle Law, Free Random Variables and Entropy* (American Mathematical Society, Providence, RI, 2000).
3. S. Thorbjørnsen, Mixed moments of Voiculescu's Gaussian random matrices, *J. Funct. Anal.* **176** (2000) 213–246.
4. G. Caire, R. R. Müller and T. Tanaka, Iterative multiuser joint decoding: Optimal power allocation and low-complexity implementation, *IEEE Trans. Inform. Theory* **50**(9) (2004) 1950–1973.
5. K. H. Fischer and J. A. Hertz, *Spin Glasses* (Cambridge University Press, Cambridge, U.K., 1991).
6. M. Mezard, G. Parisi and M. A. Virasoro, *Spin Glass Theory and Beyond* (World Scientific, Singapore, 1987).
7. T. Tanaka, A statistical mechanics approach to large-system analysis of CDMA multiuser detectors, *IEEE Trans. Inform. Theory* **48**(11) (2002) 2888–2910.
8. T. Tanaka and D. Saad, A statistical-mechanics analysis of coded CDMA with regular LDPC codes, in *Proc. of IEEE International Symposium on Information Theory (ISIT)*, Yokohama, Japan (June/July 2003), p. 444.
9. D. Guo and S. Verdú, Randomly spread CDMA: Asymptotics via statistical physics, *IEEE Trans. Inform. Theory* **51**(6) (2005) 1983–2010.
10. R. R. Müller and W. H. Gerstacker, On the capacity loss due to separation of detection and decoding, *IEEE Trans. Inform. Theory* **50**(8) (2004) 1769–1778.
11. R. R. Müller, Channel capacity and minimum probability of error in large dual antenna array systems with binary modulation, *IEEE Trans. Signal Process.* **51**(11) (2003) 2821–2828.
12. T. Tanaka and M. Okada, Approximate belief propagation, density evolution, and statistical neurodynamics for CDMA multiuser detection, *IEEE Trans. Inform. Theory* **51**(2) (2005) 700–706.
13. Y. Kabashima, A CDMA multiuser detection algorithm on the basis of belief propagation, *J. Phys. A* **36** (2003) 11111–11121.
14. H. Li and H. Vincent Poor, Impact of channel estimation errors on multiuser detection via the replica method, *EURASIP J. Wireless Commun. Networking* **2005**(2) (2005) 175–186.

15. D. Guo, Performance of multicarrier CDMA in frequency-selective fading via statistical physics, *IEEE Trans. Inform. Theory* **52**(4) (2006) 1765–1774.
16. C.-K. Wen and K.-K. Wong, Asymptotic analysis of spatially correlated MIMO multiple-access channels with arbitrary signaling inputs for joint and separate decoding, submitted to *IEEE Trans. Inform. Theory* (2004).
17. S. Verdú, *Multiuser Detection* (Cambridge University Press, New York, 1998).
18. S. V. Hanly and D. N. C. Tse, Resource pooling and effective bandwidth in CDMA networks with multiuser receivers and spatial diversity, *IEEE Trans. Inform. Theory* **47**(4) (2001) 1328–1351.
19. V. L. Girko, *Theory of Random Determinants* (Kluwer Academic Publishers, Dordrecht, The Netherlands, 1990).
20. D. N. C. Tse and S. Verdú, Optimum asymptotic multiuser efficiency of randomly spread CDMA, *IEEE Trans. Inform. Theory* **46**(7) (2000) 2718–2722.
21. D. Agrawal and A. Vardy, The turbo decoding algorithm and its phase trajectories, *IEEE Trans. Inform. Theory* **47**(2) (2001) 699–722.
22. J. J. Hopfield, Neural networks and physical systems with emergent collective computational abilities, *Proc. Natl. Acad. Sci. USA* **79** (1982) 2554–2558.
23. G. I. Kechriotis and E. S. Manolakos, Hopfield neural network implementation of the optimal CDMA multiuser detector, *IEEE Trans. Neural Networks* **7**(1) (1996) 131–141.